Lecture Notes in Mathematics 1906

Thomas Schuster

The Method of Approximate Inverse: Theory and Applications

Springer

Author

Thomas Schuster

Department of Mechanical Engineering
Helmut Schmidt University
Holstenhofweg 85
22043 Hamburg
Germany
e-mail: schuster@hsu-hh.de

Library of Congress Control Number: 2007922352

Mathematics Subject Classification (2000): 15A29, 35R30, 45Q05, 65J22, 65N21

ISSN print edition: 0075-8434
ISSN electronic edition: 1617-9692
ISBN-10 3-540-71226-7 Springer Berlin Heidelberg New York
ISBN-13 978-3-540-71226-8 Springer Berlin Heidelberg New York

DOI 10.1007/978-3-540-71227-5

Springer is a part of Springer Science+Business Media
springer.com

Typesetting by the author and SPi using a Springer LaTeX macro package
Cover design: WMXDesign GmbH, Heidelberg

Printed on acid-free paper SPIN: 12027460 VA41/3100/SPi 5 4 3 2 1 0

Dedicated to Petra, for her patience, understanding, and love

Preface

Many questions and applications in natural science, engineering, industry or medical imaging lead to inverse problems, that is: given some measured data one tries to recover a searched for quantity. These problems are of growing interest in all these disciplines and thus there is a great need for modern and stable solvers for these problems. A prominent example of an inverse problem is the problem of computerized tomography: From measured X-ray attenuation coefficients one has to calculate densities in human tissue. Mathematically inverse problems often are described as operator equations of first kind

$$\mathbf{A}f = g\,, \tag{0.1}$$

where $\mathbf{A} : X \to Y$ is a bounded operator acting on appropriate topological spaces X and Y. In case of 2D computerized tomography the mapping $\mathbf{A}$ is given by the Radon transform. Typically these operators have unbounded inverses $\mathbf{A}^{-1}$, if they are invertible at all. For instance if $\mathbf{A}$ is compact with infinite dimensional range, then $\mathbf{A}^{-1}$ is not continuous. In case of Hilbert spaces X and Y the generalized inverse $\mathbf{A}^{\dagger}$ exists and has a dense domain. But $\mathbf{A}^{\dagger}$ is bounded if and only if the range of $\mathbf{A}$ is closed which is not satisfied for compact $\mathbf{A}$. In applications the exact data g is noise contaminated e.g. by the measurement process or discretization errors. Noisy data g^{ε} lead to an useless solution $f^{\varepsilon} = \mathbf{A}^{-1}g^{\varepsilon}$ or $f^{\varepsilon} = \mathbf{A}^{\dagger}g^{\varepsilon}$ in the sense that the error $f - f^{\varepsilon}$ is unacceptably large. Hence, the stable solution of equations like (0.1) with noisy right-hand side g^{ε} require regularization methods R_{γ}. The mappings R_{γ} are bounded operators which converge pointwise to the unbounded generalized inverse $\mathbf{A}^{\dagger}$. Many regularization techniques have been developed over the last decades such as the truncated singular value decomposition, the Tikhonov-Phillips regularization or iterative methods such as the Landweber method and the method of conjugate gradients (CG-method) to name only the most popular ones.

A powerful tool which subsumes a whole family of regularization techniques is the method of approximate inverse. This method uses the duality of the operator and the spaces where it acts on. It calculates approximations to the exact solution by smoothing it with mollifiers which are approximations to Dirac's delta distribution and attenuate high frequencies contained in the solution. The method consists then of the evaluation of the measured data with so called reconstruction kernels. The reconstruction kernels themselves are solutions of an equation involving the dual operator and the mollifier and can be precomputed before the measurement process starts. A further feature of the method is its flexibility: it can be adjusted to the operator and the underlying spaces to improve the efficiency. The first idea of solving linear operator equations by mollifier methods arose in 1990 by LOUIS AND MAASS [71] and it was Louis, who published its first fundamental properties [66] and showed its regularization property [68]. RIEDER AND SCHUSTER [101, 102] derived a setting of the method for operators between arbitrary Hilbert spaces and proved convergence with rates and stability. An extension of the method to spaces of distributions was done by SCHUSTER, QUINTO [115]. The article SCHÖPFER ET AL. [107] must be seen as a first step to realize this technique in Banach spaces.

This monograph contains a comprehensive outline of the theoretical aspects of the method of approximate inverse (Part I) as well as applications of the method to different inverse problems arising in medical imaging and non-destructive testing (Parts II-IV). Part I gives a brief introduction to inverse problems and regularization methods and introduces then the approximate inverse on spaces of square integrable functions, where the Radon transform serves as a first example. We then go one step further and present the abstract setup of solving semi-discrete operator equations between arbitrary Hilbert spaces by the method of approximate inverse. Semi-discrete operator equations are of wide interest since in practical applications only a finite number of measured data is available. Part I ends with an extension of the theory to spaces of distributions. Part II puts life into the theoretical considerations of Part I and demonstrates their transfer to the problem of 3D Doppler tomography. Doppler tomography belongs to the area of medical imaging and means the problem of recovering the velocity field of a moving fluid from ultrasonic Doppler measurements. It is outlined how the method of approximate inverse leads to a solver of filtered backprojection type on the one hand and can be involved in the construction of defect correction methods on the other hand. In SONAR (SOund in NAvigation and Radiation) and SAR (Synthetic Aperture Radar) the problem arises of inverting a spherical mean operator. If the center set consists of a hyperplane this operator can no longer be described as a bounded mapping between Hilbert or Banach spaces, but it extends to a linear, continuous mapping between spaces of tempered distributions. Part III of the book presents the extension of the method to distribution spaces and shows its performance when being applied to the spherical mean operator. Further applications such as X-ray diffractometry, which is a sort of non-destructive testing, thermoacoustic tomography, where the spherical mean operator is involved, too, but with spheres as center sets, and 3D computerized tomography are the contents of Part IV. The book contains plenty of numerical results which prove that the method is well suited to cope with inverse problems in practical situations and each part is completed by a conclusion and future perspectives.

This monograph is an extended version of my habilitation thesis which I submitted at the Saarland University Saarbrücken (Germany) in 2004. The mathematical results contained therein would have been impossible to accomplish without some important people accompanying my scientific way now for many years. Thus, the first person I would like to thank is my teacher Prof. Dr. A.K. Louis who introduced me to the area of approximate inverse many years ago and who supported me all the time. Part II of the book was the result of an intensive collaboration with Prof. Dr. A. Rieder between 2000 and 2004 and I am still thankful for conveying his rich experience in approximation theory to me. Part III of the book was the result of an one year stay at Tufts University in Medford (USA) at the chair of Prof. Dr. E.T. Quinto, an acclaimed expert in integral geometry and numerical mathematics. I owe him many useful pointers with respect to the extension of the approximate inverse to distribution spaces and I will never forget his hospitality. I am further indebted to Dr. R. Müller for a very careful review of the manuscript.

Thomas Schuster Hamburg, January 2007

Contents

Part I

Inverse and Semi-discrete Problems

Many applications in natural science, industry, medicine and engineering can be concisely described by an operator equation of first kind

$$\mathbf{A}f = g\,. \tag{0.2}$$

Here, g can be seen as a set of measurement data and f is the quantity we are searching for. The mapping $\mathbf{A}$ tells how f and the data g are connected to each other. The properties of $\mathbf{A}$ have influence on the mathematical solution of (0.2). A problem like (0.2), that is observe g and find the solution f, is called an *inverse problem*.

The method of approximate inverse is a powerful and versatile tool for solving inverse problems. Articles as e.g. LOUIS [67], LOUIS, ABDULLAH [1], JONAS, LOUIS [49], LOUIS, SCHUSTER [75], RIEDER, SCHUSTER [102, 103], SCHUSTER [110] and SCHUSTER, QUINTO [115] prove this. The idea is to compute a smoothing f_γ of f. If f is a function this can be done by convolving f with e_γ which is a smooth function such as the Gaussian kernel. Then, using the duality of $\mathbf{A}$ and $\mathbf{A}^*$, f_γ can be calculated by evaluating inner products of the measured data g with a so-called reconstruction kernel v_γ which is performed in an efficient and stable way. The approximate inverse represents a class of *regularization methods*, which by its flexibility allows an adjustment to the underlying problem at a high level. Further we will see how invariance properties of $\mathbf{A}$ enhance the efficacy of the resulting inversion scheme. In practical situations equation (0.2) often is not an appropriate setting. Instead of the 'complete' function g we maybe have only a finite number of observations, e.g. moments or point evaluations, of g at hand. Thus, a semi-discrete setting

$$\mathbf{A}_n f = g_n$$

is more suited to describe real-world problems. Here, $\mathbf{A}_n$ emerges from $\mathbf{A}$ and g_n from g by an application of the so-called *observation operator* which models the measurement procedure and contains all information about the measurement device, e.g. its geometry. In order to prove strong convergence we prefer to investigate the semi-discrete rather than a fully discrete setting, though the latter one seems to be more useful from a practical point of view. Given finitely many computed moments $\langle f, e_i\rangle$ of f we obtain an approximation to f using an appropriate interpolation mapping. All these issues are subject of the first part of the book.

Part I includes five chapters. The first chapter provides essential facts from the theory of regularization methods for inverse problems. The basic idea of the approximate inverse is then demonstrated for linear, bounded mappings between L^2-spaces in the second chapter along with the two-dimensional Radon transform as a first application. In Chapter 3 we extend the method to semi-discrete problems in arbitrary Hilbert spaces. We show that the approximate inverse in fact is a regularization method and present a rigorous convergence and stability analysis. Chapter 4 finally consists of the presentation of a framework for solving semi-discrete equations in distribution spaces, which has important applications in SAR and SONAR. We furthermore sketch the idea of an error analysis for this case, too. We finish this part with some remarks collected in Chapter 5.

1

Ill-posed problems and regularization methods

We start by presenting the essential concepts for regularizing ill-posed operator equations of first kind. We refer the reader who is interested in a comprehensive treatise of this subject to the standard textbooks of LOUIS [65], ENGL, HANKE AND NEUBAUER [26], RIEDER [99], TARANTOLA [125] and HOFMANN [46][1].

We only consider the case of linear and bounded operators $\mathbf{A}$ between Hilbert spaces X and Y. The set of linear and bounded operators between X and Y is denoted by $\mathcal{L}(X,Y)$. Our aim is to investigate the solution of

$$\mathbf{A}f = g\,, \tag{1.1}$$

where $g \in Y$ is a given set of data and $f \in X$ is the quantity we want to determine. In case that $\mathbf{A}$ has a bounded inverse $\mathbf{A}^{-1}$ we obtain f by simply calculating $f = \mathbf{A}^{-1}g$. In this situation we call (1.1) *well-posed*. Unfortunately in many real applications $\mathbf{A}$ is not invertible at all. Even if $\mathbf{A}^{-1}$ exists, then the inverse might not be bounded, e.g. when $\mathbf{A}$ is compact with infinite dimensional range. Moreover, in real world problems the exact data g might not be available, but only a noise contaminated set of measurements $g^\varepsilon \in Y$ with

$$\|g^\varepsilon - g\|_Y = \varepsilon\,.$$

The corresponding solution $f^\varepsilon = \mathbf{A}^{-1}g^\varepsilon$ then usually does not converge to f if $\varepsilon \to 0$ and the defect

$$\|f^\varepsilon - f\|_X$$

can be tremendously large. The solution of equations like (1.1) where $\mathbf{A}$ has no bounded inverse $\mathbf{A}^{-1}$ is called an *ill-posed problem* due to HADAMARD.

Since often $\mathbf{A}^{-1}$ does not exist, the aim is to generalize what we understand by a *solution* of an equation like (1.1). To this end consider the defect

$$d(f) = \|\mathbf{A}f - g\|_Y\,,$$

[1] References [26], [125] are in English.

which satisfies

$$d(f)^2 = \|\mathbf{A}f - \mathcal{P}_{\overline{\mathsf{R}(\mathbf{A})}}g\|_Y^2 + \|\mathcal{P}_{\overline{\mathsf{R}(\mathbf{A})}}g - g\|_Y^2 . \tag{1.2}$$

In (1.2) $\mathcal{P}_{\overline{\mathsf{R}(\mathbf{A})}} : Y \to Y$ denotes the orthogonal projection onto $\overline{\mathsf{R}(\mathbf{A})}$, the closure of the range of $\mathbf{A}$. If $g \in \mathsf{R}(\mathbf{A}) \oplus \mathsf{R}(\mathbf{A})^{\perp}$[2], then the equation

$$\mathbf{A}f = \mathcal{P}_{\overline{\mathsf{R}(\mathbf{A})}}g \tag{1.3}$$

has a solution. From (1.2) we further deduce that every solution from (1.3) minimizes the defect $d(f)$ in X. If $\mathbf{A}$ is not injective and f a solution from (1.3), then $f + f_0$ also solves (1.3) for all f_0 in the null space $\mathsf{N}(\mathbf{A})$ of $\mathbf{A}$. As a consequence there exists a solution of (1.3) with minimal norm. This minimum norm solution lies in $\mathsf{N}(\mathbf{A})^{\perp}$ and serves as a generalized solution of $\mathbf{A}f = g$.

After defining the generalized inverse $\mathbf{A}^{\dagger}$ of a mapping $\mathbf{A}$ we note several properties of $\mathbf{A}^{\dagger}$ (Lemma 1.2). Since $\mathbf{A}^{\dagger}$ is only bounded if $\mathsf{R}(\mathbf{A})$ is closed, we need bounded approximations to $\mathbf{A}^{\dagger}$ called *regularization methods* (Definition 1.3).

Definition 1.1. *The mapping* $\mathbf{A}^{\dagger} : \mathsf{R}(\mathbf{A}) \oplus \mathsf{R}(\mathbf{A})^{\perp} \to X$*, which assigns to each element* $g \in \mathsf{R}(\mathbf{A}) \oplus \mathsf{R}(\mathbf{A})^{\perp}$ *the unique solution* $f^{\dagger}$ *of* (1.3) *in* $\mathsf{N}(\mathbf{A})^{\perp}$*, is called* generalized inverse *(Moore-Penrose inverse) of* $\mathbf{A}$.

If $\mathbf{A}$ is injective then we obviously have that $\mathbf{A}^{-1}$ and $\mathbf{A}^{\dagger}$ coincide on $\mathsf{R}(\mathbf{A})$. The following lemma contains some fundamental properties of the generalized inverse.

Lemma 1.2. *Let* $\mathbf{A} : X \to Y$ *be linear and bounded and* $g \in \mathsf{D}(\mathbf{A}^{\dagger}) := \mathsf{R}(\mathbf{A}) \oplus \mathsf{R}(\mathbf{A})^{\perp}$*. Then,*

1. $\mathbf{A}^{\dagger}$ *is linear.*
2. $f^{\dagger} = \mathbf{A}^{\dagger}g$ *is the unique solution of*

$$\mathbf{A}^* \mathbf{A}f = \mathbf{A}^* g \tag{1.4}$$

in $\mathsf{N}(\mathbf{A})^{\perp}$*, that means the unique solution of* (1.4) *with minimal norm.*

3. $\mathsf{N}(\mathbf{A}^{\dagger}) = \mathsf{R}(\mathbf{A})^{\perp}, \quad \mathsf{R}(\mathbf{A}^{\dagger}) = \mathsf{N}(\mathbf{A})^{\perp}$.
4. *The generalized inverse* $\mathbf{A}^{\dagger}$ *is bounded if and only if the renge* $\mathsf{R}(\mathbf{A})$ *of* $\mathbf{A}$ *is closed.*

Lemma 1.2 describes $f^{\dagger}$ as solution of the normal equation which is solvable whenever $g \in \mathsf{D}(\mathbf{A}^{\dagger})$. Note that $\mathsf{D}(\mathbf{A}^{\dagger}) = \mathsf{R}(\mathbf{A}) \oplus \mathsf{R}(\mathbf{A})^{\perp}$ is dense in Y. On the other hand the bad news are that $\mathbf{A}^{\dagger}$ is discontinuous if $\mathsf{R}(\mathbf{A})$ is not closed in Y which is the case if $\mathbf{A}$ has infinite dimensional range and is compact. If we

[2] For a set M in a Hilbert space the symbol $M^{\perp}$ always denotes the set of all vectors that are orthogonal to M with respect to the given inner product.

accepted $f^\dagger = \mathbf{A}^\dagger g$ as solution of (1.1) we would still run into the problem that noisy data g^ε would cause large reconstruction errors in $\mathbf{A}^\dagger g^\varepsilon$. To overcome this difficulty we approximate $\mathbf{A}^\dagger$ pointwise on its domain $\mathsf{D}(\mathbf{A}^\dagger)$ by a sequence of bounded operators. This leads us to the concept of *regularization methods.*

Definition 1.3. *Let $\mathbf{A} : X \to Y$ be bounded. A family of bounded operators $R_\gamma : Y \to X$, $\gamma \in (0, +\infty)$ is a* regularization method *for $\mathbf{A}^\dagger$, if there exists a parameter choice rule $\gamma = \gamma(\varepsilon, g^\varepsilon)$ fulfilling*

$$\lim_{\varepsilon \to 0} \sup\{\gamma(\varepsilon, g^\varepsilon) : g^\varepsilon \in Y \,,\ \|g^\varepsilon - g\|_Y < \varepsilon\} = 0 \,,$$

such that

$$\lim_{\varepsilon \to 0} \sup\{\|R_{\gamma(\varepsilon, g^\varepsilon)} g^\varepsilon - \mathbf{A}^\dagger g\|_X : g^\varepsilon \in Y \,,\ \|g^\varepsilon - g\|_Y < \varepsilon\} = 0 \tag{1.5}$$

holds true. The value γ is called regularization parameter.

We have not postulated that the family $\{R_\gamma\}$ consists of linear mappings, thus also non-linear operators are admitted. Furthermore, we only can demand the pointwise convergence of R_γ since the uniform convergence would imply the boundedness of the limit $\mathbf{A}^\dagger$ due to the Banach-Steinhaus theorem. As a consequence we have that

$$\lim_{\gamma \to 0} \|R_\gamma\|_{Y \to X} = +\infty \,.$$

Example 1.4. a) (Truncated Singular Value Decomposition)
An important tool for constructing regularization methods is the *singular value decomposition* (SVD) of a compact operator $\mathbf{A}$. The SVD consists of a sequence of triples $(\sigma_n, v_n, u_n)_{n \in \mathbb{N}} \subset \mathbb{R}_+ \times X \times Y$ with the property

$$\mathbf{A} v_n = \sigma_n \, u_n \,, \qquad \mathbf{A}^* u_n = \sigma_n \, v_n \,, \qquad n \in \mathbb{N} \,.$$

The sets $\{v_n\}_n \subset X$ and $\{u_n\}_n \subset Y$ further form complete orthonormal systems of $\mathsf{N}(\mathbf{A})^\perp = \overline{\mathsf{R}(\mathbf{A}^*)}$ and $\overline{\mathsf{R}(\mathbf{A})} = \mathsf{N}(\mathbf{A}^*)$, respectively. The sequence $\{\sigma_n\}_n$ has the unique limit point 0. The generalized inverse $\mathbf{A}^\dagger$ can be represented with the SVD as

$$\mathbf{A}^\dagger g = \sum_{\sigma_n > 0} \sigma_n^{-1} \, \langle g, u_n \rangle_Y \, v_n \,, \quad g \in \mathsf{D}(\mathbf{A}^\dagger) \,. \tag{1.6}$$

For a $g \in Y$ the *truncated SVD* is then defined by

$$R_\gamma g = \sum_{\sigma_n \geq \gamma} \sigma_n^{-1} \, \langle g, u_n \rangle_Y \, v_n \,. \tag{1.7}$$

From (1.7) and (1.6) we easily obtain the pointwise convergence to $\mathbf{A}^\dagger$.

b) (Tikhonov-Phillips regularization)
The *Tikhonov-Phillips regularization* reads as

$$R_\gamma g = (\mathbf{A}^* \mathbf{A} + \gamma^2 I)^{-1} \mathbf{A}^* g\,, \quad g \in Y\,,$$

and is equivalent to the minimization problem

$$\arg\min\{\|\mathbf{A}f - g\|_Y^2 + \gamma^2 \|f\|_X^2 : f \in X\}^3\,, \tag{1.8}$$

in the sense that $R_\gamma g$ minimizes the functional in (1.8). The penalty term $\|f\|_X^2$ can be alternated. The Tikhonov-Phillips regularization can also be written by means of the SVD as

$$R_\gamma g = \sum_{\sigma_n > 0} (\sigma_n^2 + \gamma^2)^{-1} \sigma_n \langle g, u_n \rangle_Y v_n\,, \quad g \in Y\,.$$

Both, the truncated SVD as well as the Tikhonov-Phillips regularization belong to the class of *filter methods*, since they are generated by introducing a filter in the representation for $\mathbf{A}^\dagger$ (1.6) which attenuates the small singular values causing numerical instabilities. We refer to LOUIS [65] for more details.

c) We can also see iterative methods like the *Landweber method* or the *conjugate gradient method* as regularization methods. Here, the regularization parameter is given by the number of iteration steps. The regularization properties of these methods have been investigated e.g. in HANKE [40, 41] and NEUBAUER [85].

Once we have chosen a specific regularization method, we are interested in the quality of the method. In other words: What is the convergence rate of the error $\|R_\gamma g^\varepsilon - \mathbf{A}^\dagger g\|_Y$ under certain conditions on $\mathbf{A}^\dagger g$? To this end we introduce the spaces

$$X_\nu = \{f \in \mathsf{N}(\mathbf{A})^\perp : f \in \mathsf{D}((\mathbf{A}^* \mathbf{A})^{-\nu/2})\} = \mathsf{R}((\mathbf{A}^* \mathbf{A})^{\nu/2})\,, \quad \nu \in \mathbb{R}\,,$$

which in fact turn into Hilbert spaces when we endow them with the inner product

$$\langle f_1, f_2 \rangle_\nu = \sum_{\sigma_n > 0} \sigma_n^{-2\nu} \langle f_1, v_n \rangle_X \langle f_2, v_n \rangle_X\,.$$

Moreover, because of $X_\mu \subset X_\nu$ for $\mu > \nu$, the family $\{X_\nu\}$ represents a Hilbert scale. The requirement $f \in X_\nu$ can be seen as a sort of regularity assumption: With growing ν the spaces X_ν contain 'smoother' elements. The worst case error of a regularization method for solving $\mathbf{A}f = g$ with noisy data g^ε under the assumption that $\|\mathbf{A}^\dagger g\|_\nu \le \rho$ is then given by

$$E_\nu(\varepsilon, \rho, R_\gamma) = \sup\{\|R_\gamma g^\varepsilon - \mathbf{A}^\dagger g\|_X : \|g^\varepsilon - g\|_Y < \varepsilon\,, \ \|\mathbf{A}^\dagger g\|_\nu \le \rho\}\,.$$

The unavoidable error for solving this problem is

[3] For a functional $\mathcal{J}$ the symbol arg min is defined as $f_* = \arg\min\{\mathcal{J}(f) : f \in X\} \Leftrightarrow \mathcal{J}(f_*) = \min\{\mathcal{J}(f) : f \in X\}$.

$$E_\nu(\varepsilon,\rho) = \inf_{T\in\mathcal{L}(Y,X)} E_\nu(\varepsilon,\rho,T)\,.$$

One can show the estimate

$$E_\nu(\varepsilon,\rho) \le \varepsilon^{\nu/(\nu+1)}\,\rho^{1/(\nu+1)}\,,$$

which is sharp. This leads to the definition of *optimal* and *order optimal* regularization methods.

Definition 1.5. *A regularization method $\{R_\gamma\}_\gamma$ is called* optimal *for ν, if for each $\varepsilon > 0$ and $\rho > 0$ there exists a parameter $\gamma = \gamma(\varepsilon,\rho)$ such that*

$$E_\nu(\varepsilon,\rho,R_\gamma) \le \varepsilon^{\nu/(\nu+1)}\,\rho^{1/(\nu+1)}$$

is valid. The method is called order optimal *for ν, if for each $\varepsilon > 0$ and $\rho > 0$ there exists a parameter $\gamma = \gamma(\varepsilon,\rho)$ and a constant $c \ge 1$ such that*

$$E_\nu(\varepsilon,\rho,R_\gamma) \le c\,\varepsilon^{\nu/(\nu+1)}\,\rho^{1/(\nu+1)}$$

holds true.

The Tikhonov-Phillips regularization is order optimal under certain assumptions on the degree of smoothness ν. Applying a certain parameter choice rule γ, the truncated SVD turns to a regularization method which is order optimal for all $\nu > 0$ but which is not optimal. For detailed investigations we again refer to Louis [65].

2

Approximate inverse in L^2-spaces

The method of approximate inverse includes a whole class of regularization schemes. The concept of these mollifier schemes has been established in a paper by LOUIS AND MAASS [71] in 1990, its essential properties can be read in LOUIS [66]. Just as in [71] we confine ourselves to linear, bounded operators between L^2-spaces. Throughout this chapter let $X = L^2(\Omega_1, \mu_1)$, $Y = L^2(\Omega_2, \mu_2)$, where $\Omega_i \subset \mathbb{R}^{n_i}$ are open, bounded domains and let μ_i be measures defined on Ω_i. This chapter is concerned with the solution of operator equations of first kind

$$\mathbf{A}f = g\,, \tag{2.1}$$

applying the method of approximate inverse, where $\mathbf{A} : L^2(\Omega_1, \mu_1) \to L^2(\Omega_2, \mu_2)$ is supposed to be linear and bounded.

In the first section we introduce the main idea of the method, prove its regularization property and point out its advantages considering the task of solving (2.1). One feature of this technique is that the computation of reconstruction kernels is done independently of the measured data g and thus is not influenced by noise. The second section demonstrates the performance of the method in case that $\mathbf{A} = \mathbf{R}$ is the two-dimensional *Radon transform*. The corresponding inverse problem of solving $\mathbf{R}f = g$ serves as mathematical model of two-dimensional *computerized tomography* (CT). Applying the approximate inverse to $\mathbf{R}$ leads to an inversion scheme of *filtered backprojection* type. This is a standard algorithm used in today's CT scanners.

2.1 The idea of approximate inverse

The main idea of approximate inverse is to calculate a smoothed version of the exact solution f of (2.1) or its generalized inverse $f^\dagger = \mathbf{A}^\dagger g$ rather than f itself. This is done by computing the moments

$$f_\gamma(y) = \langle f, e_\gamma(\cdot, y)\rangle_X \tag{2.2}$$

of f with an explicitly given function $e_\gamma(x,y) \in L^2(\mathbb{R}^{n_1} \times \mathbb{R}^{n_1}, \mu_1 \times \mu_2)$, $\gamma \in (0,+\infty)$. The function e_γ is to be chosen such that $e_\gamma(x,y)$ approximates *Diracs delta distribution*[1] $\delta(y-x)$ in a way being specified in Definition 2.1. In this situation we obviously have that $f_\gamma(y) \approx f(y)$, i.e. f_γ is an approximation to the original f. In general we further will choose e_γ as a smooth, at least differentiable, function implying f_γ to be smooth, too. For this reason we call e_γ a *mollifier*, that means a 'smoother'. We now specify this terminology. Note that we have $L^2(\Omega_1,\mu_1) \hookrightarrow L^2(\mathbb{R}^{n_1},\mu_1)$ putting $f(x)=0$ for $x \notin \Omega_1$. We will use this embedding without stating it explicitly.

Definition 2.1. *Let for all $y \in \mathbb{R}^{n_1}$ be $e_\gamma(\cdot,y)$ a function in $L^2(\mathbb{R}^{n_1},\mu_1)$ which has mean value equal to one for all $\gamma > 0$,*

$$\int_{\mathbb{R}^{n_1}} e_\gamma(x,y)\,\mathrm{d}\mu_1(x) = 1\,, \qquad y \in \mathbb{R}^{n_1}\,.$$

If further

$$f_\gamma(y) = \int_{\mathbb{R}^{n_1}} f(x)\,e_\gamma(x,y)\,d\mu_1(x)\,, \qquad y \in \mathbb{R}^{n_1}$$

converges to f in $L^2(\Omega_1,\mu_1)$ as $\gamma \to 0$, then we call e_γ a mollifier.

Example 2.2. Mollifiers are commonly generated by means of translation (with respect to y) and dilation (with respect to γ) of a square integrable function with normalized mean value. More specifically, take a smooth $e \in L^2(\mathbb{R}^{n_1},\mu_1)$ with $\int e(x)\,\mathrm{d}x = 1$ and set

$$e_\gamma(x,y) = \gamma^{-n_1}\, e((x-y)/\gamma)\,. \tag{2.3}$$

It is readily verified that e_γ defined by (2.3) represents a mollifier in the sense of Definition 2.1. Typical examples for functions e of such type are given by the *Gaussian function*

$$e_{\mathrm{G}}(x) = (2\,\pi)^{-n_1/2}\,\exp(-\|x\|^2/2)\,, \tag{2.4}$$

or by

$$e^\nu(x) = \kappa_\nu \begin{cases} (1-\|x\|^2)^\nu\,, & \|x\| \le 1\,, \\ 0\,, & \|x\| > 1 \end{cases} \tag{2.5}$$

where $\kappa_\nu = (\int_{\|x\|\le 1}(1-\|x\|^2)^\nu\,\mathrm{d}x)^{-1}$. The Gaussian function is arbitrarily differentiable, but does not have a compact support. The regularity of e^ν grows with ν (see [102]) and generates a mollifier $e_\gamma(\cdot,y)$ having compact support in $\{x : \|x-y\| \le \gamma\}$. The radial parts of both functions are displayed in figure 2.1, where we took the Lebesgue measure as μ_1 in these examples.
We further mention that scaling functions of wavelets with mean value one are also suited as mollifiers, see e.g. Louis, Maass and Rieder [72]. Note

[1] For continuous f we have $\delta(f) = \int f(t)\delta(t)\,\mathrm{d}t = f(0)$.

that mollifiers do not need to satisfy any symmetries though both examples (2.4) as well as (2.5) are radially symmetric. Radial symmetric mollifiers are convenient if one does not want to prefer a specific direction when smoothing f.

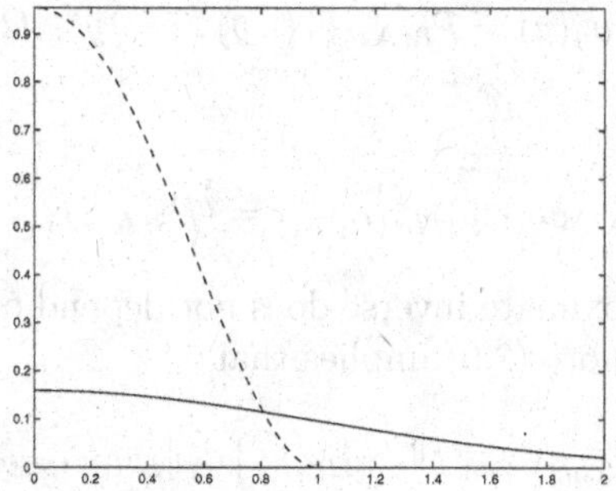

Fig. 2.1. Radial parts of the Gaussian function e_G (2.4) (solid line) and of the function e^ν (2.5) for $\nu = 2$ (dashed line)

The problem of calculating f_γ is that the unknown exact solution f appears in Definition (2.2). To overcome this difficulty we assume for the moment, that the mollifier $e_\gamma(\cdot, y)$ is in the range of the adjoint $\mathbf{A}^*$ for all $y \in \Omega_1$. Hence, equation

$$\mathbf{A}^* v_\gamma(y) = e_\gamma(\cdot, y), \qquad y \in \Omega_1 \tag{2.6}$$

has a solution $v_\gamma(y) \in Y$ and f_γ computes as

$$f_\gamma(y) = \langle f, \mathbf{A}^* v_\gamma(y)\rangle_X = \langle \mathbf{A}f, v_\gamma(y)\rangle_Y = \langle g, v_\gamma(y)\rangle_Y, \quad y \in \Omega_1 .$$

That means that the computation of f_γ can be done by a simple evaluation of inner products of the given measured data $g \in Y$ and solutions $v_\gamma(y)$ of (2.6). In general equation (2.6) is not solvable, especially if $\mathbf{A}$ is not injective and hence the range of $\mathbf{A}^*$ is not dense in X. If at least $e_\gamma(\cdot, y) \in \mathsf{D}((\mathbf{A}^*)^\dagger) = \mathsf{R}(\mathbf{A}^*) \oplus \mathsf{R}(\mathbf{A}^*)^\perp$, then the normal equation

$$\mathbf{A}\,\mathbf{A}^* v_\gamma(y) = \mathbf{A}e_\gamma(\cdot, y), \qquad y \in \Omega_1 \tag{2.7}$$

has a solution which due to Lemma 1.2 minimizes the defect

$$\min\{\|\mathbf{A}^* v - e_\gamma(\cdot, y)\|_X : v \in Y\}.$$

The mapping $g \mapsto \langle g, v_\gamma(y)\rangle_Y$ is called the *approximate inverse* of $\mathbf{A}$.

Definition 2.3. *Suppose* $\mathbf{A} : L^2(\Omega_1, \mu_1) \to L^2(\Omega_2, \mu_2)$ *to be linear and bounded and let* e_γ *be a mollifier satisfying* $e_\gamma(\cdot, y) \in \mathsf{D}((\mathbf{A}^*)^\dagger)$ *for* $y \in \Omega_1$. *The linear mapping* $\widetilde{\mathbf{A}}_\gamma : L^2(\Omega_2, \mu_2) \to L^2(\Omega_1, \mu_1)$ *defined by*

$$\widetilde{\mathbf{A}}_\gamma g(y) = \langle g, v_\gamma(y) \rangle_{L^2(\Omega_2,\mu_2)}\,, \quad y \in \Omega_1\,, \tag{2.8}$$

where $v_\gamma(y)$ *solves* (2.7) *is called the* (continuous) approximate inverse *of* $\mathbf{A}$. *A solution* $v_\gamma(y)$ *of* (2.7) *is the* reconstruction kernel *associated to* e_γ.

Lemma 1.2 tells us that every solution of (2.7) validates

$$\mathbf{A}^* v_\gamma(y) = \mathcal{P}_{\mathsf{R}(\mathbf{A}^*)} e_\gamma(\cdot, y)\,, \qquad y \in \Omega_1\,.$$

Thus, we have that

$$\widetilde{\mathbf{A}}_\gamma \mathbf{A} f(y) = \langle f, \mathcal{P}_{\mathsf{R}(\mathbf{A}^*)} e_\gamma(\cdot, y) \rangle_{L^2(\Omega_1,\mu_1)} = \langle \mathcal{P}_{\mathsf{R}(\mathbf{A}^*)} f, e_\gamma(\cdot, y) \rangle_{L^2(\Omega_1,\mu_1)} \tag{2.9}$$

proving that the approximate inverse does not depend on the particular solution of (2.7). Furthermore (2.9) implies that

$$\|\widetilde{\mathbf{A}}_\gamma g\|_{L^2(\Omega_1,\mu_1)} \le \|f\|_{L^2(\Omega_1,\mu_1)}\, \|e_\gamma\|_{L^2(\mathbb{R}^{n_1}\times\mathbb{R}^{n_2},\mu_1\times\mu_2)}\,,$$

from what we deduce that $\widetilde{\mathbf{A}}_\gamma g$ is in $L^2(\Omega_1, \mu_1)$ and the operator $\widetilde{\mathbf{A}}_\gamma$ is well-defined. The intention of introducing the approximate inverse was to get a class of regularization methods. And in fact we can prove pointwise convergence of $\widetilde{\mathbf{A}}_\gamma g$ to $\mathbf{A}^\dagger g$.

Theorem 2.4. *The operator* $\widetilde{\mathbf{A}}_\gamma : Y \to X$ *is continuous, if*

$$l_\gamma := \left(\int_{\Omega_1} \|v_\gamma(y)\|^2_{L^2(\Omega_2,\mu_2)}\, \mathrm{d}\mu_1(y) \right)^{1/2} < \infty\,. \tag{2.10}$$

In case that (2.10) *is satisfied, then* $\|\widetilde{\mathbf{A}}_\gamma\|_{Y\to X} \le l_\gamma$.
Further, if $g = \mathbf{A} f$, $g \in \mathsf{D}(\mathbf{A}^\dagger)$, *then the convergence*

$$\lim_{\gamma\to 0} \widetilde{\mathbf{A}}_\gamma g = \mathbf{A}^\dagger g\,, \quad g \in \mathsf{D}(\mathbf{A}^\dagger) \tag{2.11}$$

holds true.

Proof. From (2.8) we immediately deduce for $g \in Y$

$$\|\widetilde{\mathbf{A}}_\gamma g\|^2_{L^2(\Omega_1,\mu_1)} = \int_{\Omega_1} \left| \langle g, v(y) \rangle_{L^2(\Omega_2,\mu_2)} \right|^2 \mathrm{d}\mu_1(y) \le l_\gamma^2\, \|g\|^2_{L^2(\Omega_2,\mu_2)}\,.$$

To prove (2.11) we use (2.9) and Definition 1.1 to obtain

$$\widetilde{\mathbf{A}}_\gamma g(y) = \langle \mathcal{P}_{\mathsf{R}(\mathbf{A}^*)} f, e_\gamma(\cdot, y) \rangle_{L^2(\Omega_1,\mu_1)} = \langle \mathbf{A}^\dagger g, e_\gamma(\cdot, y) \rangle_{L^2(\Omega_1,\mu_1)}\,.$$

Since e_γ has the mollifier property according to Definition 2.1, the convergence (2.11) is verified. □

Remark 2.5. To prove stability with respect to noise we would have to state a parameter choice rule for γ depending of the noise level ε in the data g^ε such that

$$\lim_{\varepsilon\to 0} \sup\{\|\widetilde{\mathbf{A}}_{\gamma(\varepsilon,g^\varepsilon)} g^\varepsilon - \mathbf{A}^\dagger g\|_X : g^\varepsilon \in Y\,,\ \|g^\varepsilon - g\|_Y < \varepsilon\} = 0$$

is guaranteed according to Definition 1.3. An application of the triangle inequality yields

$$\begin{aligned}\|\widetilde{\mathbf{A}}_\gamma g^\varepsilon - \mathbf{A}^\dagger g\|_X &\le \|\widetilde{\mathbf{A}}_\gamma\|_{Y\to X}\, \|g^\varepsilon - g\|_Y + \|\widetilde{\mathbf{A}}_\gamma g - \mathbf{A}^\dagger g\|_X \\ &\le l_\gamma\, \varepsilon + \|\widetilde{\mathbf{A}}_\gamma g - \mathbf{A}^\dagger g\|_X\,.\end{aligned}$$

Because of (2.11) we obtain convergence to 0 of the second summand for every sequence $\{\gamma\}$ tending to 0. The sequence $\{l_\gamma\}$ however in general diverges as $\gamma \to 0$. Hence, we have to find a parameter choice rule $\{\gamma = \gamma(\varepsilon)\}$ tending to 0 in such a way that

$$\lim_{\varepsilon\to 0} l_{\gamma(\varepsilon)}\, \varepsilon = 0\,.$$

That procedure, which is typical for regularizing ill-posed problems, will be outlined in detail in Part II when applying the method to the problem of *Doppler tomography.*

So far we did not say anything about how to compute reconstruction kernels, that means how to solve equation (2.6) or (2.7), respectively. We point out two possibilities: one by means of an inversion formula for $\mathbf{A}$, the second with the help of the SVD.

Assume that $\mathbf{A}$ is injective, $e_\gamma(\cdot, y) \in \mathsf{R}(\mathbf{A}^*)$ and that we have an inversion formula

$$f = \mathbf{B}\,\mathbf{A}f \tag{2.12}$$

at our disposal, where $\mathbf{B} : Y \to X$ is bounded. Then, a solution of (2.6) is given by

$$v_\gamma(y) = \mathbf{B}^* e_\gamma(\cdot, y) \tag{2.13}$$

with $\mathbf{B}^* : X \to Y$ the adjoint of $\mathbf{B}$. Since $\mathbf{A}^*\,\mathbf{B}^* = I_{|\mathsf{R}(\mathbf{A}^*)}$, $v_\gamma(y)$ is in fact a reconstruction kernel for $\mathbf{A}$.

If $\mathbf{A}$ is compact, then there exists a SVD $\{\sigma_n, v_n, u_n\}_{n\in\mathbb{N}}$. The unique solution of (2.7) in $\mathsf{N}(\mathbf{A}^*)^\perp$, that is $v_\gamma(y) = (\mathbf{A}^*)^\dagger e_\gamma(\cdot, y)$, is then represented according to (1.6) by

$$v_\gamma(y) = \sum_{\sigma_n > 0} \sigma_n^{-1} \langle e_\gamma(\cdot, y), v_n\rangle_{L^2(\Omega_1,\mu_1)}\, u_n \tag{2.14a}$$

$$= \sum_{\sigma_n > 0} \sigma_n^{-2} \langle \mathbf{A} e_\gamma(\cdot, y), u_n\rangle_{L^2(\Omega_2,\mu_2)}\, u_n\,. \tag{2.14b}$$

In practical situations we have to cut off the series (2.14b) after finitely many steps leading to unwanted artifacts in the reconstructions. That is why the cut-off index must be chosen carefully.

A further possibility to get an approximate solution of (2.6) is to use a *projection method*. To do so we have to define subspaces $X_h \subset X$ and $Y_h \subset Y$ of finite dimension and then to seek for a solution $v_\gamma^h \in Y_h$ of

$$\mathcal{P}_h \, \mathbf{A}^* v_\gamma^h = \mathcal{P}_h e_\gamma(\cdot, y)\,. \tag{2.15}$$

With $\mathcal{P}_h : X \to X_h$ we denote the orthogonal projection onto X_h. Expanding

$$v_\gamma^h(y) = \sum_{j=1}^{\dim Y_h} \alpha_j \, \phi_j$$

in a basis $\{\phi_j\}$ of Y_h transforms (2.15) into a system of linear equations with the coefficients $\{\alpha_j\}$ as solution. This is then to be solved by an exact or iterative inversion scheme. A projection method to get reconstruction kernels has been applied in [111], where the author has used a collocation method to compute kernels for the *Laplace transform*. That was necessary since there is neither an inversion formula nor the SVD available for the Laplace transform in $\mathbb{R}$. For a thorough study of projection methods we highly recommend NATTERER'S article [79].

The fact that the equation (2.6) does not depend on the set of data g^ε and thus the computation of reconstruction kernels is independent of noise must be seen as a great advantage of the method of approximate inverse. But the method is not efficient in the present form, since (2.6) has to be solved for all reconstruction points $y \in \Omega_1$, i.e. all values y at which we want to recover f. Even in two dimensions this would be much too time consuming, making the method not applicable to large scale problems even though the calculating of the kernels can be done *before* the measurement process starts. There is a possibility to overcome that predicament, if the operator $\mathbf{A}$ satisfies certain invariance properties.

Theorem 2.6. *Let $T_1^y : X \to X$, $T_2^y : Y \to Y$ be bounded operators on X and Y for $y \in \Omega_1$, respectively, satisfying*

$$T_2^y \, \mathbf{A}\,\mathbf{A}^* = \mathbf{A}\,\mathbf{A}^* \, T_2^y\,, \quad T_2^y \, \mathbf{A} = \mathbf{A} \, T_1^y\,, \qquad y \in \Omega_1\,.$$

Further assume that the mollifier e_γ is generated by T_1^y. That means, there exists a $y^\star \in \Omega_1$ with

$$e_\gamma(\cdot, y) = T_1^y e_\gamma(\cdot, y^\star)\,, \qquad y \in \Omega_1\,.$$

If $v_\gamma(y^\star)$ is a solution of $\mathbf{A}\,\mathbf{A}^* v_\gamma(y^\star) = \mathbf{A} e_\gamma(\cdot, y^\star)$, then

$$v_\gamma(y) = T_2^y v_\gamma(y^\star) \tag{2.16}$$

solves (2.7) for arbitrary $y \in \Omega_1$.

Proof. The assertion follows from

$$\begin{aligned} \mathbf{A}e_\gamma(\cdot,y) &= \mathbf{A}\,T_1^y e_\gamma(\cdot,y^\star) = T_2^y\,\mathbf{A}e_\gamma(\cdot,y^\star) \\ &= T_2^y\,\mathbf{A}\,\mathbf{A}^* v_\gamma(y^\star) = \mathbf{A}\,\mathbf{A}^*\,T_2^y v_\gamma(y^\star)\,. \end{aligned}$$

□

Theorem 2.6 has been taken from LOUIS [66]. In case of an injective operator $\mathbf{A}$ an invariance property as $T_1^y\,\mathbf{A}^* = \mathbf{A}^*\,T_2^y$ for $y \in \Omega_1$ is sufficient to verify (2.16). Since the normal equation (2.7) is equivalent to

$$\mathbf{A}^* v_\gamma(y) = \mathcal{P}_{\mathsf{R}(\mathbf{A}^*)} e_\gamma(\cdot,y)\,,$$

the conjecture comes up that an intertwining with respect to $\mathbf{A}^*$ is enough even in more general situations. Actually, we will prove a theorem with weaker assumptions in Part II. Note that $\mathbf{A}$ might not satisfy any invariance properties at all. In that case we can not apply Theorem 2.6.

Thanks to the invariance property (2.16) we only have to solve the normal equation (2.7) for a *single* reconstruction point, namely y^*. The remaining reconstruction kernels are then produced using $v_\gamma = T_2^y v_\gamma(y^\star)$. This procedure saves a lot of computation time. The method of (continuous) approximate inverse finally reads

$$\widetilde{\mathbf{A}}_\gamma g(y) = \langle g, T_2^y v_\gamma(y^\star)\rangle_{L^2(\Omega_2,\mu_2)}\,, \qquad y \in \Omega_1\,.$$

The invariance mappings T_i^y, $i = 1,2$, in most cases rely on symmetry properties of the underlying measurement geometries.

2.2 A first example: The Radon transform

A norm $\|\cdot\|$ without subscript always denotes the Euclidean norm in $\mathbb{R}^n$, $\|x\| = \|x\|_2 = \sqrt{\langle x,x\rangle_2}$. The *Radon transform* assigns a function f defined on $\mathbb{R}^2$, or part of it, to its integrals over all lines. We confine to functions defined on the unit disk in $\mathbb{R}^2$. More explicitly, let $\Omega^2 = B_1(0) = \{x \in \mathbb{R}^2 : \|x\| < 1\}$ be the open unit disk in $\mathbb{R}^2$ and $Z = [0,\pi] \times [-1,1]$. A line L intersecting Ω^2 is determined by the polar angle $\varphi \in [0,\pi]$ of its normal and its distance from the origin $s \in [-1,1]$,

$$L(\varphi,s) = \{x \in \mathbb{R}^2 : \langle x, \omega(\varphi)\rangle = s\}\,, \qquad \varphi \in [0,\pi]\,,\; s \in [-1,1]\,.$$

Here, $\omega(\varphi) = (\cos\varphi, \sin\varphi)^\top \in S^1 = \partial\Omega^2$ is the unit normal vector to $L(\varphi,s)$. Hence, the vector $\omega(\varphi)^\perp = (-\sin\varphi, \cos\varphi)$, being perpendicular to ω, is a vector of direction associated to $L(\varphi,s)$. The situation is emphasized in figure 2.2.

For $f \in L^2(\Omega^2)$, the Radon transform is then defined by

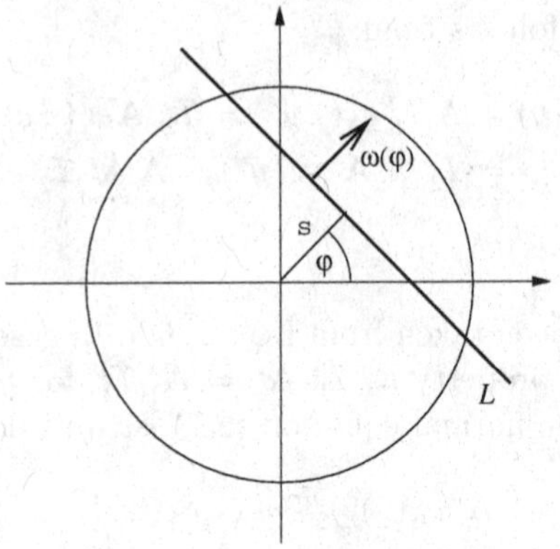

Fig. 2.2. Parameters of the Radon transform

$$\mathbf{R}f(\varphi, s) = \int_{L(\varphi,s)\cap\Omega^2} f(x)\, \mathrm{d}\ell(x)\,. \tag{2.17}$$

Since the publication of RADON's [95] fundamental article, where he also stated an inversion formula for $\mathbf{R}$, the mathematical properties as well as the development of inversion schemes for that mapping have been object of thorough research. The reason is that, amongst other applications, the Radon transform represents the mathematical model of the two-dimensional *computerized tomography* (CT). The results of this research can e.g. be found in NATTERER [80], NATTERER, WÜBBELING [84], KAK, SLANEY [51], or HELGASON [43]. At first we summarize the essential mathematical properties of $\mathbf{R}$.

By virtue of the mapping (2.17) $\mathbf{R}$ is a linear, bounded operator between $L^2(\Omega^2)$ and $L^2(Z)$. The corresponding adjoint $\mathbf{R}^* : L^2(Z) \to L^2(\Omega^2)$ has the representation

$$\mathbf{R}^* g(x) = \int_0^\pi g(\omega(\varphi), \langle x, \omega(\varphi)\rangle)\, \mathrm{d}\varphi \tag{2.18}$$

and integrates a function g over all lines intersecting x. That is why $\mathbf{R}$ is called *backprojection.* The Radon transform is injective and an inversion formula is given by

$$f = \frac{1}{2\pi} \mathbf{R}^* \Lambda \mathbf{R} f\,, \qquad f \in L^2(\Omega^2)\,, \tag{2.19}$$

where $\mathbf{F}\,\Lambda g(\varphi, \sigma) = |\sigma|\, \mathbf{F} g(\varphi, \sigma)$ denotes the *Riesz potential* and

$$\mathbf{F} g(\varphi, \sigma) = (2\pi)^{-1/2} \int_{\mathbb{R}} g(\varphi, s)\, \mathrm{e}^{-\mathrm{i}\,\sigma\, s}\, \mathrm{d}s$$

is the one-dimensional *Fourier transform* with respect to the variable s.

We give an outline how to determine reconstruction kernels for $\mathbf{R}$ by means of formula (2.19) and the usage of appropriate invariances due to Theorem 2.6. The kernels are then be used to formulate the method of approximate inverse from Section 2.1 to solve the inverse problem

$$\mathbf{R}f = g. \tag{2.20}$$

Equation (2.20) can be interpreted as the mathematical formulation of the two-dimensional CT problem: We have to reconstruct the density function f from measured line integrals g.

Let $e \in L^2(\mathbb{R}^2)$ be a function with mean value 1,

$$\int_{\mathbb{R}^2} e(x)\,\mathrm{d}x = 1.$$

We generate a mollifier by translating and dialating e,

$$e_\gamma(x,y) = T_{1,\gamma}^y e(x) = \gamma^{-2}\, e\Big(\frac{x-y}{\gamma}\Big), \qquad x, y \in \mathbb{R}^2. \tag{2.21}$$

It is easily verified that

$$\lim_{\gamma\to 0} \langle f, e_\gamma(\cdot, y)\rangle_{L^2(\mathbb{R}^2)} = f(y)$$

in $L^2(\mathbb{R}^2)$ and hence that e_γ is a mollifier in the sense of Definition (2.1). Furthermore $T_{1,\gamma}^y : L^2(\mathbb{R}^2) \to L^2(\mathbb{R}^2)$ is linear, continuous and satisfies an intertwining with the backprojection $\mathbf{R}^*$.

Lemma 2.7. *Fix $y \in \mathbb{R}^2$ and define*

$$T_{2,\gamma}^y g(\varphi, s) = \gamma^{-2}\, g\Big(\varphi, \frac{s - \langle y, \omega(\varphi)\rangle}{\gamma}\Big), \qquad g \in L^2((0,\pi)\times\mathbb{R}).$$

Then, $T_{2,\gamma}^y$ is linear and bounded on $L^2((0,\pi)\times\mathbb{R})$ and obeys the invariance property

$$T_{1,\gamma}^y\, \mathbf{R}^* = \mathbf{R}^*\, T_{2,\gamma}^y. \tag{2.22}$$

Proof. Since $T_{2,\gamma}^y$ is a composition of linear, bounded operators, the first part of Lemma 2.7 is obvious. Assertion (2.22) follows from

$$\begin{aligned}
\mathbf{R}^*\, T_{2,\gamma}^y g(x) &= \gamma^{-2} \int_0^\pi g\Big(\varphi, \frac{\langle x, \omega(\varphi)\rangle - \langle y, \omega(\varphi)\rangle}{\gamma}\Big)\,\mathrm{d}\varphi \\
&= \gamma^{-2} \int_0^\pi g\Big(\varphi, \frac{\langle x - y, \omega(\varphi)\rangle}{\gamma}\Big)\,\mathrm{d}\varphi \\
&= \gamma^{-2}\mathbf{R}^* g\Big(\frac{x-y}{\gamma}\Big) = T_{1,\gamma}^y\, \mathbf{R}^* g(x).
\end{aligned}$$

□

With the help of (2.22) we can show how to produce an arbitrary kernel $\upsilon_\gamma(y)$ associated to $e_\gamma(\cdot, y)$ from a single one.

Corollary 2.8. *Suppose* $e \in \mathsf{R}(\mathbf{R}^*)$ *and* υ *to be a solution of*

$$\mathbf{R}^* \upsilon = e . \tag{2.23}$$

Then,

$$\upsilon_\gamma(y) = \mathcal{T}^y_{2,\gamma} \upsilon \tag{2.24}$$

solves $\mathbf{R}^* \upsilon_\gamma(y) = e_\gamma(\cdot, y)$, *where* e_γ *is the mollifier* (2.21).

Proof. This corollary is an immediate consequence from (2.22), the injectivity of $\mathbf{R}$ and Theorem 2.6. □

Corollary 2.8 tells us that we only have to solve (2.23). The kernels $\upsilon_\gamma(y)$ are then generated by applying $\mathcal{T}^y_{2,\gamma}$ to that solution. Note that (2.23) is equivalent to the normal equation, since $\mathbf{R}$ is injective. A solution υ of (2.23) can be obtained using the inversion formula (2.19). We get

$$\upsilon = \frac{1}{2\,\pi} \Lambda \mathbf{R} e . \tag{2.25}$$

To calculate υ from (2.25) explicitly we use the *Fourier slice theorem.* For $f \in L^2(\mathbb{R}^2)$ we have, that

$$\mathbf{F}\mathbf{R} f(\varphi, \sigma) = (2\,\pi)^{1/2}\, \hat{f}(\sigma\,\omega(\varphi)), \qquad \sigma \in \mathbb{R}, \quad \varphi \in [0, \pi] . \tag{2.26}$$

On the right-hand side of (2.26) we have used the two-dimensional Fourier transform

$$\mathbf{F} f(\xi) = \hat{f}(\xi) = (2\,\pi)^{-1} \int_{\mathbb{R}^2} f(x)\, \mathrm{e}^{-\mathrm{i}\,\langle \xi, x\rangle}\, \mathrm{d}x .$$

Since we do not want to prefer a particular direction when filtering f, we choose a mollifier e which is radially symmetric, $e(x) = e(\|x\|)$.

Lemma 2.9. *Let* $e \in L^2(\mathbb{R}^2)$ *be radially symmetric with* $\int_{\mathbb{R}^2} e(x)\, \mathrm{d}x = 1$. *Then,*

$$\upsilon(s) = \pi^{-1} \int_0^\infty \sigma\, \hat{e}(\sigma\,\omega(0))\, \cos(s\,\sigma)\, \mathrm{d}\sigma \tag{2.27}$$

solves (2.23). *Particularly, the solution is independent of* φ.

Proof. Using (2.25) and the Fourier slice Theorem (2.26) we get

$$\begin{aligned}\hat{\upsilon}(\varphi, \sigma) &= (2\,\pi)^{-1}\, \mathbf{F} \Lambda \mathbf{R} e(\varphi, \sigma) = (2\,\pi)^{-1}\, |\sigma|\, \mathbf{F}\mathbf{R} e(\varphi, \sigma) \\ &= (2\,\pi)^{-1/2}\, |\sigma|\, \hat{e}(\sigma\,\omega(\varphi)) .\end{aligned}$$

Since e is radially symmetric, its Fourier transform $\hat{e}$ also does not depend on the direction and we may write $\hat{e}(\sigma\,\omega(\varphi)) = \hat{e}(\sigma\,\omega(0))$. An application of the inverse Fourier transform finally yields

$$\begin{aligned} \upsilon(s) &= (2\,\pi)^{-1} \int_{-\infty}^{\infty} |\sigma|\,\hat{e}(\sigma\,\omega(0))\,\mathrm{e}^{\mathrm{i}\,s\,\sigma}\,\mathrm{d}\sigma \\ &= \pi^{-1} \int_0^{\infty} \sigma\,\hat{e}(\sigma\,\omega(0))\,\cos(s\,\sigma)\,\mathrm{d}\sigma\,. \end{aligned}$$

□

Example 2.10. a) The Gaussian (2.4) is radially symmetric with mean value equal to 1. Taking into account $\hat{e_{\mathrm{G}}}(\sigma) = (2\,\pi)^{-1}\,\exp(-\sigma^2/2)$, the corresponding reconstruction kernel (for $n_1 = 2$) computes as

$$\begin{aligned} \upsilon_{\mathrm{G}}(s) &= \frac{1}{2\,\pi^2} \int_0^{\infty} \sigma\,\exp(-\sigma^2/2)\,\cos(s\,\sigma)\,\mathrm{d}\sigma \\ &= -\frac{1}{2\,\pi^2} \int_0^{\infty} \frac{\partial}{\partial\sigma}\Big(\exp(-\sigma^2/2)\Big)\,\cos(s\,\sigma)\,\mathrm{d}\sigma\,. \end{aligned}$$

Applying formulae (7.4.7) and (7.1.3) from ABRAMOWITZ, STEGUN [2] we obtain

$$\upsilon_{\mathrm{G}}(s) = \frac{1}{2\,\pi^2}\left(1 + \mathrm{i}\,\sqrt{\frac{\pi}{2}}\;s\,\exp(-s^2/2)\,\mathrm{erf}(\mathrm{i}\,s/\sqrt{2})\right), \tag{2.28}$$

where $\mathrm{erf}(t) = (2/\sqrt{\pi})\int_0^t \exp(-z^2)\,\mathrm{d}z$ denotes the *error function*. Just as e_{G} the reconstruction kernel υ_{G} does not have a compact support.

b) The function (2.5) represents a radially symmetric mollifier with compact support but without the smoothness of the Gaussian. Let $n_1 = 2$, and $\nu \in \mathbb{N}$. Then, e^ν reads

$$e^\nu(x) = \frac{\nu+1}{\pi}\begin{cases} (1 - \|x\|^2)^\nu\,, & \|x\| \le 1\,, \\ 0\,, & \|x\| > 1\,. \end{cases} \tag{2.29}$$

The factor $(\nu+1)/\pi$ was chosen such that $\int e^\nu(x)\,\mathrm{d}x = 1$. The smoothness of e^ν increases with ν, we have $e^\nu \in H_0^\alpha(\Omega^2)$ whenever $\alpha < \nu + 1/2$. We are interested in the reconstruction kernel υ^ν associated with e^ν. Using spherical coordinates we compute

$$\hat{e^\nu}(\sigma\,\omega(0)) = \frac{\nu+1}{\pi}\int_0^1 r\,(1-r^2)^\nu\,\mathrm{J}_0(r\,\sigma)\,\mathrm{d}r\,, \tag{2.30}$$

where

$$J_\mu(z) = \left(\frac{z}{2}\right)^\mu \sum_{k=0}^{\infty} \frac{(-1)^k}{k!\,\Gamma(\mu+k+1)} \left(\frac{z}{2}\right)^{2k} \tag{2.31}$$

is the *Bessel function* of first kind and order μ and Γ denotes *Eulers gamma function.* Putting the series expansion (2.31) into (2.30) and switching the order of summation and integration leads to

$$\begin{aligned} \hat{e^\nu}(\sigma\,\omega(0)) &= \frac{(\nu+1)!}{2\,\pi} \sum_{k=0}^{\infty} (-1)^k \frac{\sigma^{2k}}{4^k\,k!\,\Gamma(\nu+k+2)} \\ &= \frac{2^\nu\,(\nu+1)!}{\pi}\,\sigma^{-(\nu+1)}\,J_{\nu+1}(\sigma)\,. \end{aligned} \tag{2.32}$$

We use again (2.27) together with formula (6.699.2) from GRADSHTEYN, RIZHIK [32] to find the representation

$$\begin{aligned} v^\nu(s) &= \frac{2^\nu\,(\nu+1)!}{\pi^2} \int_0^\infty \sigma^{-\nu}\,J_{\nu+1}(\sigma)\,\cos(s\,\sigma)\,d\sigma \\ &= \frac{1}{2\,\pi^2} \begin{cases} 2\,(\nu+1)\,{}_2F_1(1,-\nu;1/2;s^2)\,, & |s| \le 1\,, \\ -s^{-2}\,{}_2F_1(1,3/2;\nu+2;s^{-2})\,, & |s| > 1\,. \end{cases} \end{aligned} \tag{2.33}$$

Here, ${}_2F_1$ means the *hypergeometric function.* Note that the kernel v^ν does not have a compact support whereas e^ν has one, namely $\overline{\Omega^2}$, see (2.29). The reason is the Riesz potential Λ which occurs in (2.25) and is a non-local operator. Certainly the smoothness of v^ν grows with ν just as it does for the mollifier e^ν. Figure 2.3 illustrates the reconstruction kernels v_G as well as v^ν for different values of ν.

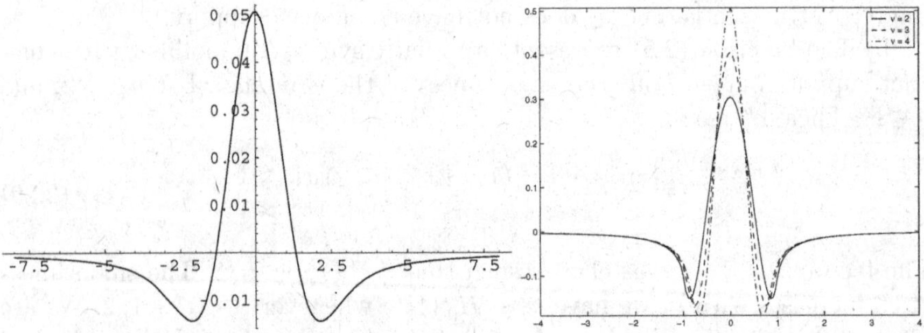

Fig. 2.3. Plot of the reconstruction kernels v_G (left picture) and v^ν for $\nu = 2, 3, 4$ (right picture)

Assume e to be a mollifier and v the corresponding reconstruction kernel, that is $\mathbf{R}^* v = e$, then the method of approximate inverse to solving (2.20) for given measured data $g \in L^2(Z)$ reads

$$\widetilde{\mathbf{R}}_\gamma g(y) = \langle g, T^y_{2,\gamma} v\rangle_{L^2(Z)}$$

$$= \gamma^{-2} \int_0^\pi \int_{-1}^1 g(\varphi, s)\, v\Big(\varphi, \frac{s - \langle y, \omega(\varphi)\rangle}{\gamma}\Big)\, \mathrm{d}s\, \mathrm{d}\varphi \tag{2.34}$$

according to Corollary 2.8. A discretization of (2.34) by applying a trapezoidal sum leads to the method of *filtered backprojection*, the most common algorithm in computerized tomography, see e.g. NATTERER [80]. The term 'filtered backprojection' is explained by the fact that the inner integration in (2.34) can be seen as a filtering of the measured data g followed by the backprojection, that is the summation over all lines intersecting the reconstruction point y. Figure 2.4 shows reconstructions of the well known Shepp-Logan head phantom using the kernels v_{G} and v^ν for $\nu = 6$. We see that the Gaussian has a large smoothing effect making the boundaries of the ellipses a little fuzzy, whereas the mollifier e^6 yields an image with contoures which are pretty visible.

One might argue that the application of the approximate inverse to the Radon inversion does not yield a novel solution scheme for this inverse problem. But the convergence and stability analysis in Section 3.2 allows for error estimates with requirements for the solution f which are significantly weaker then previous ones. Furthermore we did not have any a priori constraints to a coupling of the discretization step size in s and the regularization parameter γ as it is the case for the widely used Shepp-Logan filter, see SHEPP, LOGAN [117] and NATTERER [80].

Remark 2.11. If $v^\Lambda_\gamma(y)$ solves $\mathbf{R}^* v^\Lambda_\gamma(y) = \Lambda e_\gamma(\cdot, y)$, then we have

$$\langle \Lambda f, e_\gamma(\cdot, y)\rangle_{L^2(\mathbb{R}^2)} = \langle \mathbf{R} f, v^\Lambda_\gamma\rangle$$

because of the symmetry of Λ. Thus, we get a smoothed version of Λf when using $v^\Lambda_\gamma(y)$ as kernel for the approximate inverse. Since the Λ-operator preserves the singular support of f, computing Λf emphasizes singularities such as edges and jumps of f. This sort of tomography is called *lambda tomography*. Standard references for lambda and local tomography are VAINBERG AND FANGOIS [127], FARIDANI ET AL. [29, 27], and RASHID-FARROKHI ET AL. [97]. The application of the method of approximate inverse to lambda tomography is outlined in RIEDER, DIETZ AND SCHUSTER [100].

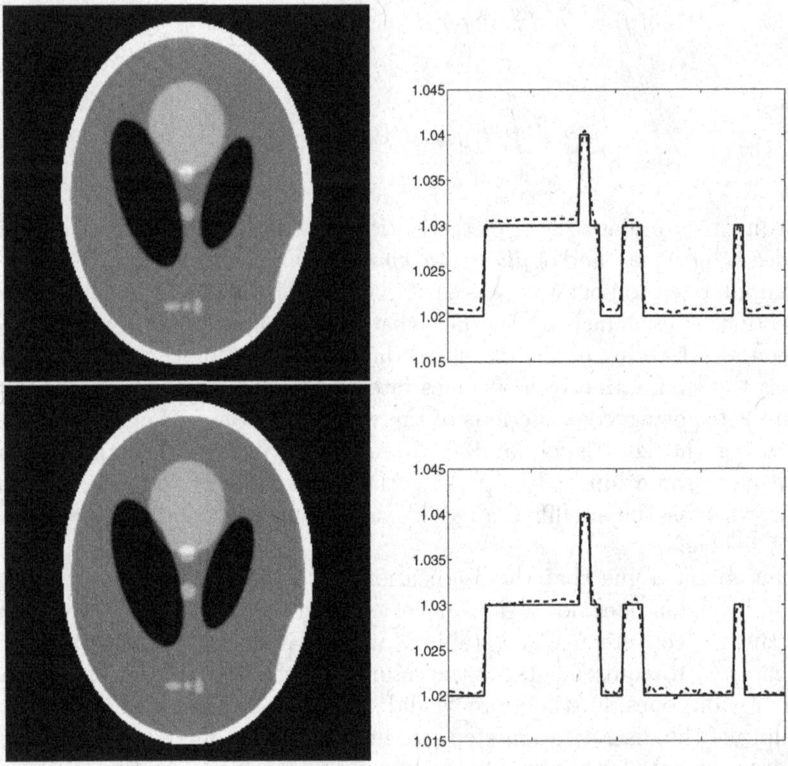

Fig. 2.4. Reconstruction of the Shepp-Logan head phantom using the Gaussian e_{G} as mollifier (top left) and e^6 (bottom left). To the right corresponding cross sections are displayed.

'Courtesy of Prof. Dr. A. Rieder, University Karlsruhe, 76123 Karlsruhe, Germany. The pictures are taken from his article *On filter design principles in 2D computerized tomography*, in Radon Transforms and Tomography, E. Q. et al., ed., vol. 278 of Contemporary Mathematics, AMS Publications, 2001, pp. 207–226. Copyright ©2001 American Mathematical Society Publications. Reproduced with permission'.

3

Approximate inverse in Hilbert spaces

We go one step further and focus at linear and bounded operators

$$\mathbf{A} : X \to Y$$

between real or complex Hilbert spaces X and Y.

The method of continuous approximate inverse introduced in Chapter 2 is not close to reality with respect to two points. On the one hand, in practical situations we only have a finite number of data available rather than a function g. On the other hand, we evaluate the inner products $\langle g, v_\gamma(y)\rangle$ for finitely many points $y \in \Omega_1$ only and not in the whole of Ω_1. This predicament suggests to consider the semi-discrete equation

$$\mathbf{A}_n f = g_n$$

instead of $\mathbf{A}f = g$. Here, $g_n \in \mathbb{K}^n$, $\mathbb{K} = \mathbb{R}, \mathbb{C}$, is the vector containing the n measurement data and $\mathbf{A}_n$ emerges from $\mathbf{A}$ by a discretization to be specified. Once we have stated the concept of a mollifier in an arbitrary Hilbert space and have a bunch of mollifers $\{e_i\}_{i=1}^d \subset X$ at hand, our aim is to approximate moments

$$\langle f, e_i\rangle_X\,, \qquad i = 1, \ldots, d$$

in a way similar to (2.2) using the finite set of data g_n only. Using the moments $\langle f, e_i\rangle_X$ we approximate f by an interpolation operator and show strong convergence in X.

Thus the idea of this chapter is not only the generalization of the approximate inverse to Hilbert spaces but also to establish a framework well suited to address real world problems.

3.1 Semi-discrete operator equations

Let $\mathbb{K}$ be the field of real or complex numbers, respectively. As mentioned we can not expect to have all data $g \in Y$ available. Moreover, we model the measurement process by a so-called *observation operator*

$$\Psi_n : Y \to \mathbb{K}^n \tag{3.1}$$

and assume that data $g_n = \Psi_n g \in \mathbb{K}^n$ are given. The observation operator contains all details about the measurement device like the measurement geometry and the particular discretization. Further we suppose Ψ_n to be continuous, e.g. we may consider g_n as a finite number of moments of g which are measured. The task is then to find an element $f \in X$ satisfying

$$\mathbf{A}_n f = g_n \,, \tag{3.2}$$

where $\mathbf{A}_n = \Psi_n \mathbf{A}$. Since problem (3.2) is underdetermined, the minimum-norm solution $f_n^\dagger = \mathbf{A}_n^\dagger g_n$, that is the unique solution of

$$\mathbf{A}_n^* \mathbf{A}_n f = \mathbf{A}_n^* g_n \tag{3.3}$$

in $\mathsf{N}(\mathbf{A}_n)^\perp$, is the best we can hope for. Since $\dim \mathsf{R}(\mathbf{A}_n) < \infty$ and hence closed, $\mathbf{A}_n^\dagger g_n$ is well defined for all $g_n \in \mathbb{K}^n$. Applying the method of approximate inverse we actually compute moments $\langle f^\dagger, e_\gamma(\cdot, y)\rangle$ of the generalized inverse with a mollifier as can be seen from equation (2.9). Thus, we might calculate moments

$$\langle f_n^\dagger, e_i \rangle_X \,, \qquad i = 1, \ldots, d \tag{3.4}$$

of the minimum-norm solution $f_n^\dagger$ with mollifiers e_i, $i = 1, \ldots, d$, if we want to extend the method to solve problems like (3.2). In case that $X = L^2(\Omega)$ such a set of mollifiers $\{e_i\}$ is e.g. given by $e_i(x) = \gamma^{-2} e((x - y_i)/\gamma)$, where e is a radial symmetric function having mean value 1 and y_i, $i = 1, \ldots, d$ represent the reconstruction points in Ω. In that situation the moments (3.4) approximate the values $f_n^\dagger(y_i)$ for finitely many y_i.

At first, we have to define what we mean by a *mollifier* in an arbitrary Hilbert space. The aim is to use an interpolation operator to approximate $f_n^\dagger$ with the help of the computed moments (3.4). More explicitly, given $\langle f_n^\dagger, e_i \rangle_X$ for $i = 1, \ldots, d$, compute

$$E_d f_n^\dagger = \sum_{i=1}^{d} \langle f_n^\dagger, e_i \rangle_X \, b_i \,, \tag{3.5}$$

where $\{b_i\}_{i=1}^d$ is a family in X associated to $\{e_i\}_{i=1}^d$ and is specified later. The mapping $E_d : X \to X$ assigns values $\langle f_n^\dagger, e_i \rangle_X$, which are obtained applying the method of approximate inverse, to elements of X and we want that the convergence

$$\lim_{d \to \infty} \| E_d f_n^\dagger - f_n^\dagger \|_X = 0$$

holds true. To this end we postulate the family $\{b_i\}_{i=1}^d$ to satisfy

$$\frac{c}{d} \sum_{i=1}^{d} |\alpha_i|^2 \le \Big\| \sum_{i=1}^{d} \alpha_i \, b_i \Big\|_X^2 \le \frac{C}{d} \sum_{i=1}^{d} |\alpha_i|^2 \,, \qquad \alpha \in \mathbb{K}^d \,, \tag{3.6}$$

where $C > c > 0$ are two constants which are independent of d. A system $\{b_i\}$, for which (3.6) is valid, is called *Riesz system*. The estimates (3.6) imply that the b_i are linearly independent. All that gives rise to the following definition.

Definition 3.1. *Let $\{e_i\}_{i=1}^d$ be a subset of X and $\{b_i\}_{i=1}^d$ be a Riesz system in X. We say that $E_d : X \to X$ with*

$$E_d f = \sum_{i=1}^{d} \langle f, e_i \rangle_X \, b_i \tag{3.7}$$

has the mollifier property, *if*

$$\lim_{d\to\infty} \|E_d f - f\|_X = 0\,, \qquad f \in X \tag{3.8}$$

is satisfied.

Note that the convergence property (3.8) implies that the families $\{e_i\}$ und $\{b_i\}$ are related to each other. The reader might be interested whether systems fulfilling (3.8) do exist. Hence, we continue by presenting a family $\{e_i\}_{i=1}^d$ which has the mollifier property.

Example 3.2. Suppose $\Omega^m = \{x \in \mathbb{R}^m : \|x\| < 1\}$ is the open unit disc in $\mathbb{R}^m$, $X = L^2(\Omega^m)$ and e to be a radially symmetric function with compact support and mean value equal to 1. Such a function is e.g. given by (2.29). For $d \in \mathbb{N}$ we define a sequence $\{e_{d,i}\}_{i\in\mathbb{Z}^m}$ in $L^2(\mathbb{R}^m)$ by $e_{d,i}(x) = d^m\, e(d\,x - i)$ where $d \geq 2$ and $i \in \mathbb{Z}^m$. If e has the support $\overline{\Omega^m} = \overline{B_1(0)}$[1], then $e_{d,i}$ is supported in $\overline{B_{d^{-1}}(d^{-1}\,i)}$. The reconstruction points y_i are then equal to $d^{-1}\,i$.

Next we have to fix a Riesz system $\{B_{d,i}\}$. Let for $t \in \mathbb{R}$ the linear B-spline be given by

$$b(t) := b^{(1)} = \begin{cases} 1 - |t|\,, & |t| \leq 1\,, \\ 0\,, & |t| > 1 \end{cases}$$

and set $b_{d,k}(t) = b(d\,t - k)$ for $k = 1, \ldots, d-1$ and

$$b_{d,0}(t) = \chi_{[0,d^{-1}]}(t)\, b(d\,t) \qquad \text{und} \qquad b_{d,d}(t) = \chi_{[1-d^{-1},1]}(t)\, b(d\,t - d)\,,$$

where $\chi_I(t)$ always denotes the *characteristic function* of an interval I. Let further $B = b \otimes b \otimes \ldots \otimes b$ be the m-fold tensor product of the B-spline b and $B_{d,i}(x) = B(d\,x - i)$. We define a mapping $E_d^{(m)} : L^2(\mathbb{R}^m) \to L^2(\mathbb{R}^m)$ by

$$E_d^{(m)} f(x) = \sum_{i\in\mathbb{Z}^m} \langle f, e_{d,i} \rangle_{L^2(\mathbb{R}^m)}\, B_{d,i}(x)\,. \tag{3.9}$$

Note that, since e is compactly supported, we have $\langle f, e_{d,i} \rangle_{L^2(\mathbb{R}^m)} \neq 0$ only for finitely many $i \in \mathbb{Z}^m$. Using $\int e(x)\,\mathrm{d}x = 1$ and $\int x_j\, e(x)\,\mathrm{d}x = 0$, $j = 1, \ldots, m$, which follows from the radial symmetry of e, we deduce that

[1] $B_R(z)$ always denotes the open ball $\{x \in \mathbb{R}^m : \|x - z\| < R\}$.

$$E_d^{(m)} p(x) = p(x)\,, \qquad x \in \mathbb{R}^m \tag{3.10}$$

where p is a polynomial in m variables with degree less than or equal to 1. Thus, the mapping $E_d^{(m)}$ reproduces polynomials of degree 1. This is essential to prove the mollifier property of $E_d^{(m)}$.

The key idea to prove (3.8) is to use an argument from BRAMBLE, HILBERT [11]. To do so we need the local boundedness of $E_d^{(m)}$ in addition to the conservation property (3.10). Let $\square = (0,1)^n$ be the n-dimensional unit cube and $\square_{d,r} = d^{-m}(\square + r)$ be a translated and dilated version of it. If we denote $\mathcal{F}_r = \{i \in \mathbb{Z}^m : \operatorname{supp} B(\cdot - i) \cap \square_{1,r}\}$, then $\mathcal{F}_r$ is finite because of the compact support of B and we obtain

$$\begin{aligned}
\|E_d^{(m)} f\|^2_{L^2(\square_{d,r})} &= \sum_{k,i \in \mathcal{F}_r} \langle f, e_{d,k}\rangle_{L^2(\square_{d,r})} \overline{\langle f, e_{d,i}\rangle_{L^2(\square_{d,r})}}\, \mathcal{K}_{k,i} \\
&\le \sum_{k \in \mathcal{F}_r} |\langle f, e_{d,k}\rangle_{L^2(\square_{d,r})}|^2 \, \|\mathcal{K}\|_2\,,
\end{aligned} \tag{3.11}$$

where the real and symmetric matrix $\mathcal{K}$ of dimension $|\mathcal{F}_r|$ is given by

$$\begin{aligned}
\mathcal{K}_{k,i} &= d^{-m} \int_{\square_{1,r}} B(x-k)\, \overline{B(x-i)}\, \mathrm{d}x \\
&= d^{-m} \prod_{j=1}^{m} \int_{r_j}^{r_j+1} b(x_j - k_j)\, \overline{b(x - i_j)}\, \mathrm{d}x_j\,.
\end{aligned}$$

Here, we used the tensor product structure of B. We proceed by estimating the spectral norm $\|\mathcal{K}\|_2$ of $\mathcal{K}$. Because $\mathcal{K}$ is symmetric, $\|\mathcal{K}\|_2$ is equal to the eigenvalue with greatest absolute value. Gershgorin's theorem says that all eigenvalues of a matrix $\mathcal{A}$ of dimension n are contained in the set

$$\bigcup_{l=1}^{n} \Big\{ \zeta \in \mathbb{C} : |\zeta - \mathcal{A}_{ll}| \le \sum_{\substack{k=1\\k\neq l}}^{n} |\mathcal{A}_{lk}| \Big\}\,,$$

see e.g. HANKE-BOURGEOIS [42, Theorem 23.1]. Since $|\mathcal{K}_{k,i}|$ is bounded by a multiple of d^{-m}, Gershgorin's theorem implies that

$$\|\mathcal{K}\|_2 \le C_B\, d^{-m}\,.$$

This together with $\|e_{d,k}\|^2_{L^2(\mathbb{R}^m)} \le C_e\, d^m$ and (3.11) leads to the estimate

$$\begin{aligned}
\|E_d^{(m)} f\|^2_{L^2(\square_{d,r})} &\le C_B\, d^{-m} \sum_{k \in \mathcal{F}_r} \|f\|^2_{L^2(\operatorname{supp}\, e_{d,k})} \|e_{d,k}\|^2_{L^2(\mathbb{R}^m)} \\
&\le C_B\, C_e \|f\|^2_{L^2(S_{d,r})}\,,
\end{aligned} \tag{3.12}$$

where $S_{d,r} = \bigcup_{k\in\mathcal{F}_r} \operatorname{supp} e_{d,k}$. Since $E_d^{(m)}$ reproduces polynomials of degree 1 we have for $p \in \Pi_1^m$[2]

$$\begin{aligned}\|E_d^{(m)} f - f\|_{L^2(\square_{d,r})} &\le \|f - p\|_{L^2(\square_{d,r})} + \|E_d^{(m)}(p - f)\|_{L^2(\square_{d,r})}\\ &\le (1 + C_B^{1/2} C_e^{1/2}) \|f - p\|_{L^2(S_{d,r})},\end{aligned}$$

and since p was arbitrary we finally get

$$\|E_d^{(m)} f - f\|_{L^2(\square_{d,r})} \le c \inf_{p\in\Pi_1^m} \|f - p\|_{L^2(S_{d,r})}.$$

Hence, the error $\|E_d^{(m)} f - f\|_{L^2(\square_{d,r})}$ can be estimated by the approximation power of polynomials of degree 1. Jackson's theorem characterizes how well a function can be approximated by polynomials of fixed degree. It can be found in SCHUMAKER [108, Theorem 3.12]. An application of that theorem yields

$$\inf_{p\in\Pi_1^m} \|f - p\|_{L^2(S_{d,r})} \le c\,\omega_\alpha(f; d^{-1})_2,$$

where $\omega_\alpha(f; d^{-1})_2$ denotes the αth modulus of smoothness of f in $L^2(S_{d,r})$. The moduli of smoothness are described in detail in Section 2.8 in the book of Schumaker [108]. There, one also finds the estimate

$$\omega_\alpha(f; t)_2 \le t^\alpha |f|_{H^\alpha(S_{d,r})}$$

which finally gives

$$\|E_d^{(m)} f - f\|_{L^2(\square_{d,r})} \le c\,d^{-\alpha} |f|_{H^\alpha(S_{d,r})}, \qquad 0 \le \alpha \le 2.$$

Here, $|\cdot|_{H^\alpha(S_{d,r})}$ denotes the H^α-seminorm on $S_{d,r}$. We conclude

$$\|E_d^{(m)} f - f\|_{L^2(\mathbb{R}^m)} \le c\,d^{-\min\{2,\alpha\}} |f|_{H^\alpha(\mathbb{R}^m)},$$

and obtain for $f \in H_0^\alpha(\Omega^m) \subset H^\alpha(\mathbb{R}^m)$

$$\|E_d^{(m)} f - f\|_{L^2(\Omega^m)} \le \|E_d^{(m)} f - f\|_{L^2(\mathbb{R}^m)} \le c\,d^{-\min\{2,\alpha\}} |f|_{H^\alpha(\Omega^m)}. \quad (3.13)$$

Since $H_0^\alpha(\Omega^m)$ is dense in $L^2(\Omega^m)$ the mollifier property

$$\lim_{d\to\infty} \|E_d^{(m)} f - f\|_{L^2(\Omega^m)} = 0, \qquad f \in L^2(\Omega^m). \quad (3.14)$$

is verified. □

[2] By Π_k^m we always denote the multivariate polynomials in m variables with degree less than or equal to k.

We have seen that for $X = L^2(\Omega^m)$ systems $\{e_i\}$, $\{b_i\}$ which validate the convergence (3.8) actually do exist. Suppose we have approximations for $\langle f_n^\dagger, e_i\rangle_X$ – in a way to be specified later – then the mollifier property yields $E_d f_n^\dagger \to f_n^\dagger$ as $d \to \infty$.

So far, we did not specify how we obtain approximations to $\langle f_n^\dagger, e_i\rangle_X$. To this end we introduce, analogously to the L^2-case in Chapter 2, reconstruction kernels v_i^n which are to minimize the defect $\|\mathbf{A}_n^* w - e_i\|_X$. In other words v_i^n solves the normal equation

$$\mathbf{A}_n\, \mathbf{A}_n^* v_i^n = \mathbf{A}_n e_i\,, \qquad i = 1, \ldots, m\,. \tag{3.15}$$

We have an analogue to (2.9) in this semi-discrete setting too.

Lemma 3.3. *If* $\mathbf{A}_n \in \mathcal{L}(X, \mathbb{K}^n)$ *and either* $g \in \mathsf{R}(\mathbf{A})$ *or* v_i^n *the unique solution of* (3.15) *in* $\mathsf{R}(\mathbf{A}_n)$, *then*

$$\langle f_n^\dagger, e_i\rangle_X = \langle g_n, v_i^n\rangle_{\mathbb{K}^n}\,, \tag{3.16}$$

where $g_n = \Psi_n g$.

Proof. From Lemma 1.2 we have $\mathbf{A}_n^* v_i^n = \mathcal{P}_{\mathsf{N}(\mathbf{A}_n)^\perp} e_i$ leading to

$$\langle f_n^\dagger, e_i\rangle_X = \langle f_n^\dagger, \mathcal{P}_{\mathsf{N}(\mathbf{A}_n)^\perp} e_i\rangle_X = \langle \mathbf{A}_n f_n^\dagger, v_i^n\rangle_{\mathbb{K}^n} = \langle \mathcal{P}_{\mathsf{R}(\mathbf{A}_n)} g_n, v_i^n\rangle_{\mathbb{K}^n}\,.$$

If $g \in \mathsf{R}(\mathbf{A})$, then $g_n = \Psi_n g = \Psi_n\, \mathbf{A} u$ for $u \in X$ and we immediately obtain $\mathcal{P}_{\mathsf{R}(\mathbf{A}_n)} g_n = g_n$. If, on the other hand, $v_i^n \in \mathsf{R}(\mathbf{A}_n)$, then obviously $\mathcal{P}_{\mathsf{R}(\mathbf{A}_n)} v_i^n = v_i^n$. In both cases (3.16) is verified. □

Equation (3.16) motivates the following definition for the approximate inverse in the semi-discrete setting for Hilbert spaces.

Definition 3.4. *Assume* $\mathbf{A}_n : X \to \mathbb{K}^n$ *to be linear and bounded. Let* $\{e_i\} \subset X$, $\{b_i\} \subset X$ *be systems satisfying the mollifier property* (3.8). *Further let* $v_i^n \in \mathbb{K}^n$, $i = 1, \ldots, d$ *be solutions of the normal equations* (3.15). *The mapping* $\widetilde{\mathbf{A}}_{n,d} : \mathbb{K}^n \to X$ *defined as*

$$\widetilde{\mathbf{A}}_{n,d} w = \sum_{i=1}^{d} \langle w, v_i^n\rangle_{\mathbb{K}^n}\, b_i \tag{3.17}$$

is called (semi-discrete) approximate inverse *of* $\mathbf{A}_n$.

Because of (3.8) and (3.16) we have the convergence

$$\widetilde{\mathbf{A}}_{n,d}\, \mathbf{A}_n f = E_d f_n^\dagger \to f_n^\dagger = \mathcal{P}_{\mathsf{N}(\mathbf{A}_n)^\perp} f \quad \text{as } d \to \infty\,. \tag{3.18}$$

So far, we translated the concepts established in Chapter 2 to the semi-discrete setting in arbitrary Hilbert spaces. The computation of the reconstruction kernels however bears two crucial difficulties:

1.) The matrix $\mathbf{A}_n \mathbf{A}_n^*$ is not sparse and has a large dimension corresponding to the number of data n. This number grows tremendously, if we consider higher-dimensional problems. Since we furthermore handle the discretization of an inverse problem we must expect that $\mathbf{A}_n \mathbf{A}_n^*$ is ill-conditioned leading to unstable solutions even when the right-hand side in (3.15) is not contaminated by noise. In that case one might use iterative solvers yielding an approximate solution only.

2.) Even worse is the situation where $\mathbf{A}_n : \mathsf{D}(\mathbf{A}_n) \subset X \to \mathbb{K}^n$ is unbounded. This situation e.g. appears when $\mathbf{A} = \mathbf{R}$ is the Radon transform and Ψ_n are point evaluations in $(\varphi_k, s_l) \in ([0,\pi] \times [-1,1]) = Z$. RIEDER, SCHUSTER [101, Appendix A] proved that

$$\mathbf{R} : H_0^\alpha(\Omega^2) \to H^{\alpha+1/2}(Z) \tag{3.19}$$

is continuous for $\alpha \geq 0$. Note that in (3.19) the smoothing by the factor $1/2$ is to be understood with respect to both variables φ and s and thus differs from known results in NATTERER [81, Chap. II, Theorem 5.3], LOUIS AND NATTERER [73] or HAHN AND QUINTO [36]. Hence, the domain $\mathsf{D}(\mathbf{R}_n)$ of $\mathbf{R}_n$ is given by $\mathsf{D}(\mathbf{R}_n) = \mathsf{D}(\Psi_n \mathbf{R}) = H_0^\alpha(\Omega^2)$ for $\alpha > 1/2$. In RIEDER, SCHUSTER [102, Theorem 5.1] the authors construct sequences of functions $\{f_k\}_{k\in\mathbb{N}}$ in $H_0^\alpha(\Omega^2)$, $\alpha > 1/2$, satisfying $\|f_k\|_{L^2(\Omega^2)} \leq 1$ but $\|\mathbf{R}_n f_k\|_{\mathbb{R}^n} \not\to \infty$ as $k \to \infty$. This proves that $\mathsf{D}(\mathbf{R}_n^*) = \{0\}$[3] and hence that the adjoint $\mathbf{R}_n^*$ does not exist. Examples where $\mathbf{A}_n$ is bounded on X are integral operators with sufficiently smooth kernels.

To include also unbounded $\mathbf{A}_n$, like the semi-discrete Radon transform $\mathbf{R}_n$, to the concept of approximate inverse, we assume that there exist Banach spaces X_1, Y_1 with continuous and dense embeddings $X_1 \hookrightarrow X$ and $Y_1 \hookrightarrow Y$ such that

$$\mathbf{A} : X_1 \to Y_1$$

is bounded. In case $\mathbf{A} = \mathbf{R}$, due to the considerations made before, such Banach spaces are $X_1 = H_0^\alpha(\Omega^2)$, $Y_1 \subset H^{\alpha+1/2}(Z)$. The observation operator $\Psi_n : Y_1 \to \mathbb{K}^n$ is assumed to be given as

$$(\Psi_n v)_k = \langle \psi_{n,k}, v \rangle_{Y_1^* \times Y_1}\,, \qquad k = 1, \ldots, n\,,$$

where $\psi_{n,k} \in Y_1^*$, $k = 1, \ldots, n$, are linear and bounded functionals on Y_1, Y_1^* is the topological dual of Y_1 and $\langle \cdot, \cdot \rangle_{Y_1^* \times Y_1}$ denotes the dual pairing for $Y_1^* \times Y_1$[4]. Obviously $\mathsf{D}(\mathbf{A}_n) = X_1$ and $\mathbf{A}_n : X_1 \to \mathbb{K}^n$ is linear. If $\mathbf{A}_n$ is bounded, then $X_1 = X$ also topologically and the reconstruction kernels v_i^n are well defined as solutions from (3.15). If $\mathbf{A}_n$ is unbounded, then $\mathbf{A}_n^*$ does

[3] For unbounded $\mathbf{A}_n$ we have that $\mathsf{D}(\mathbf{A}_n^*) = \{v \in \mathbb{K}^n : f \mapsto \langle \mathbf{A}_n f, v \rangle_{\mathbb{K}^n}$ is continuous on $\mathsf{D}(\mathbf{A}_n)\}$, see e.g. RUDIN [105, Chapter 13].

[4] The topological dual Y^* of a Banach space Y consists of all linear and bounded functionals $y^* : Y \to \mathbb{K}$. The dual pairing is given by $y^*(y) =: \langle y^*, y \rangle_{Y^* \times Y}$ for $y \in Y$.

not even need to exist as in case of the Radon transform. In this situation, the reconstruction kernels are not meaningfully defined by (3.15).

If $\mathbf{A}_n : X_1 \to \mathbb{K}^n$ is unbounded, we suggest the following procedure. Let $\epsilon_i > 0$, $i = 1, \dots, d$, be given. Then there exist v_i in Y_1 with

$$\|\mathcal{P}_{\mathsf{N}(\mathbf{A})^\perp} e_i - \mathbf{A}^* v_i\|_X \leq \epsilon_i \,, \qquad i = 1, \dots, d \tag{3.20}$$

because Y_1 is dense in Y. Such 'continuous' kernels can be computed by means of the singular value decomposition of $\mathbf{A}$ or an inversion formula as outlined in Section 2.1. If $\mathbf{A}$ is injective, then we can even postulate $\epsilon_i = 0$ as we have seen in Section 2.2. As a replacement for the non-existing discrete kernels v_i^n we essentially take the vectors which arise when applying the observation operator Ψ_n to v_i. More explicitly,

$$v_i^n = G_n \Psi_n v_i \,, \qquad i = 1, \dots, d \,. \tag{3.21}$$

Here, $G_n \in \mathbb{K}^{n \times n}$ is the *Gramian matrix* of a Riesz system $\{\phi_k\}_{k=1}^n \subset Y$ which is connected with Ψ_n and will define an interpolation operator $\Pi_n : Y_1 \to Y$. The Gramian matrix G_n then is given as

$$(G_n)_{k,l} = \langle \phi_k, \phi_l \rangle_Y \,, \qquad 1 \leq k, l \leq n \,.$$

We will specify the connection of the family $\{\phi_k\}$ with Ψ_n as well as the interpolation operator Π_n in Section 3.2. Note, that the application of Ψ_n to v_i is well defined because of $v_i \in Y_1$. The kernels (3.21) can then be used to formulate the method of approximate inverse $\widetilde{\mathbf{A}}_{n,d}$ according to (3.17).

3.2 Convergence and stability

The question arises to what extent the convergence (3.18) is satisfied and, if it does not longer hold true, which additional constraints we have to include that it is still valid. Furthermore by now we did not take into consideration the influence of measurement errors in the data g_n. All that is subject of this section.

As mentioned, we aim to generate kernels with the help of the observation operator Ψ_n by (3.21) and use them as replacements to define the semi-discrete approximate inverse $\widetilde{\mathbf{A}}_{n,d}$ by (3.17). In view of convergence results it is necessary to assign the discrete values $\Psi_n v_i$ to an element of Y. This is done by means of an interpolation operator $\Pi_n : Y_1 \to Y$, which has to obey two fundamental conditions: one is a boundedness, the other an approximation condition. To this end let $\{\phi_k\}_{k=1}^n \subset Y$ be a Riesz system in Y. The mapping $\Pi_n : Y_1 \to Y$ is defined by

$$\Pi_n v = \sum_{k=1}^n (\Psi_n v)_k \, \phi_k = \sum_{k=1}^n \langle \psi_{n,k}, v \rangle_{Y_1^* \times Y_1} \, \phi_k \,, \quad v \in Y_1 \,. \tag{3.22}$$

Just as in the case of the mollifier operator E_d, the choice of $\{\phi_k\}_{k=1}^n$ is not arbitrary, either. Moreover the family $\{\phi_k\}_{k=1}^n$ has to be such that the two mentioned conditions are satisfied which we will specify now.

On the one hand Π_n has to fulfill an *approximation property.* Assume $\{\rho_n\}_n \subset [0,1]$ to be a monotonically decreasing sequence tending to zero such that

$$\|\Pi_n v - v\|_Y \leq C_\Pi\, \rho_n\, \|v\|_{Y_1}\,, \qquad v \in Y_1\,, n \to \infty \tag{3.23}$$

with $C_\Pi > 0$ independent from n.

On the other hand we postulate *uniform boundedness* from Π_n

$$\|\Pi_n\|_{Y_1 \to Y} \leq C_b \qquad \text{for } n \to \infty\,, \tag{3.24}$$

where the constant $C_b > 0$ does not depend on n either.

Postulating the conditions (3.23) and (3.24), it is clear that the choice of the system $\{\phi_k\}$ depends on the observation operator Ψ_n. Imagine that Ψ_n represents point evaluations of a function, then Π_n in fact is an interpolation operator.

If we denote by $G_n \in \mathbb{K}^n$ the *Gramian matrix* with respect to $\{\phi_k\}$, i.e. $(G_n)_{k,l} = \langle \phi_k, \phi_l \rangle_Y$, then

$$\langle \Psi_n v, G_n\, \Psi_n w \rangle_{\mathbb{K}^n} = \langle \Pi_n v, \Pi_n w \rangle_Y\,, \qquad v, w \in Y_1\,. \tag{3.25}$$

Equation (3.25) provides an important relation between Π_n and Ψ_n. Note that the matrix G_n also appears in the Definition (3.21) of the replacement kernels v_i^n.

Lemma 3.5. *There exists a constant $c > 0$ which does not depend on n such that the estimate*

$$|\langle v, w \rangle_Y - \langle \Pi_n v, \Pi_n w \rangle_Y| \leq c\, \rho_n\, \|v\|_{Y_1}\, \|w\|_{Y_1} \qquad v, w \in Y_1 \tag{3.26}$$

is valid for $n \to \infty$.

Proof. Using the approximation property (3.23), the boundedness (3.24) and the triangle inequality we obtain

$$\begin{aligned} |\langle v, w \rangle_Y - \langle \Pi_n v, \Pi_n w \rangle_Y| &\leq |\langle v - \Pi_n v, w \rangle_Y| + |\langle \Pi_n v, w - \Pi_n w \rangle_Y| \\ &\leq C_\Pi\, C_{Y,Y_1}\, \rho_n\, \|v\|_{Y_1}\, \|w\|_{Y_1} + C_\Pi\, C_b\, \rho_n\, \|v\|_{Y_1}\, \|w\|_{Y_1}\,, \end{aligned}$$

where C_{Y,Y_1} is the norm of the continuous embedding $Y_1 \hookrightarrow Y$,

$$\sup_{\|y\|_{Y_1}=1} \|y\|_Y = C_{Y,Y_1}\,.$$

The assertion follows then with $c := C_\Pi\,(C_{Y,Y_1} + C_b)$. □

We have all ingredients together to formulate the convergence statement for $\widetilde{\mathbf{A}}_{n,d}$.

Theorem 3.6. *Let the mappings* $\mathbf{A}$, E_d, Ψ_n *and* Π_n *be given as in sections* 3.1 *and* 3.2. *Further assume that the triples*

$$\{(e_i, v_i, b_i)\}_{i=1}^d \subset X \times Y_1 \times X$$

satisfy (3.8), (3.20) *and that* $\widetilde{\mathbf{A}}_{n,d}$ *is defined by* (3.17) *where* $v_i^n = G_n\,\Psi_n v_i$, $i = 1, \ldots, d$. *If* $f \in X_1$, *then there exists a constant* $C > 0$ *which does not depend on* d *and* n, *such that*

$$\|\widetilde{\mathbf{A}}_{n,d}\,\mathbf{A}_n f - \mathcal{P}_{\mathsf{N}(\mathbf{A})^\perp} f\|_X \le \|(I - E_d)\,\mathcal{P}_{\mathsf{N}(\mathbf{A})^\perp} f\|_X \qquad (3.27)$$
$$+C\left(\frac{1}{d}\sum_{i=1}^d (\rho_n^2\,\|v_i\|_{Y_1}^2 + \epsilon_i^2)\right)^{1/2} \|f\|_{X_1}\,.$$

If further $d^{-1}\sum_{i=1}^d \epsilon_i^2 \to 0$ *for* $d \to \infty$ *and* $\rho_n^2\, d^{-1}\sum_{i=1}^d \|v_i\|_{Y_1}^2 \to 0$ *as* $n, d \to \infty$, *then we have convergence*

$$\lim_{\substack{n\to\infty \\ d\to\infty}} \|\widetilde{\mathbf{A}}_{n,d}\,\mathbf{A}_n f - \mathcal{P}_{\mathsf{N}(\mathbf{A})^\perp} f\|_X = 0\,, \qquad f \in X_1\,.$$

Proof. By means of the triangle inequality we estimate

$$\|\widetilde{\mathbf{A}}_{n,d}\,\mathbf{A}_n f - \mathcal{P}_{\mathsf{N}(\mathbf{A})^\perp} f\|_X \le \|(I - E_d)\,\mathcal{P}_{\mathsf{N}(\mathbf{A})^\perp} f\|_X + \|E_d\,\mathcal{P}_{\mathsf{N}(\mathbf{A})^\perp} f - \widetilde{\mathbf{A}}_{n,d}\,\mathbf{A}_n f\|_X\,.$$

Using the identity

$$\langle \mathcal{P}_{\mathsf{N}(\mathbf{A})^\perp} f, e_i\rangle_X = \langle f, \mathcal{P}_{\mathsf{N}(\mathbf{A})^\perp} e_i\rangle_X = \langle f, \mathcal{P}_{\mathsf{N}(\mathbf{A})^\perp} e_i - \mathbf{A}^* v_i\rangle_X + \langle \mathbf{A} f, v_i\rangle_X$$

and applying (3.20) and (3.26) yields

$$\begin{aligned}
&\|E_d\,\mathcal{P}_{\mathsf{N}(\mathbf{A})^\perp} f - \widetilde{\mathbf{A}}_{n,d}\,\mathbf{A}_n f\|_X^2 \\
&= \Big\|\sum_{i=1}^d \Big(\langle \mathcal{P}_{\mathsf{N}(\mathbf{A})^\perp} f, e_i\rangle_X - \langle \mathbf{A}_n f, G_n\,\Psi_n v_i\rangle_{\mathbb{K}^n}\Big)\, b_i\Big\|_X^2 \\
&\le \frac{1}{d}\sum_{i=1}^d |\langle \mathcal{P}_{\mathsf{N}(\mathbf{A})^\perp} f, e_i\rangle_X - \langle \Psi_n\,\mathbf{A} f, G_n\,\Psi_n v_i\rangle_{\mathbb{K}^n}|^2 \\
&\le \frac{4}{d}\sum_{i=1}^d \{|\langle f, \mathcal{P}_{\mathsf{N}(\mathbf{A})^\perp} e_i - \mathbf{A}^* v_i\rangle_X|^2 + |\langle \mathbf{A} f, v_i\rangle_Y - \langle \Pi_n\,\mathbf{A} f, \Pi_n v_i\rangle_Y|^2\} \\
&\le \frac{4}{d}\sum_{i=1}^d \{\epsilon_i^2\,\|f\|_X^2 + c^2\,\rho_n^2\,\|\mathbf{A} f\|_{Y_1}^2\,\|v_i\|_{Y_1}^2\} \\
&\le \frac{4}{d}\sum_{i=1}^d \{C_{X,X_1}^2\,\epsilon_i^2 + c^2\,\rho_n^2\,\|\mathbf{A}\|_{X_1\to Y_1}^2\,\|v_i\|_{Y_1}^2\}\|f\|_{X_1}^2\,,
\end{aligned}$$

where $C_{X,X_1} > 0$ is the norm of the continuous embedding $X_1 \hookrightarrow X$. Setting $C := \max\{2\,C_{X,X_1}, 2\,c\,\|\mathbf{A}\|_{X_1\to Y_1}\}$ with c from (3.26) completes the proof. □

Theorem 3.6 shows that convergence of the semi-discrete approximate inverse of $\mathbf{A}_n$ to the part of f being perpendicular to the null space of $\mathbf{A}$ actually is possible, if we only have a suitable coupling of ρ_n and ϵ_i. Though in Theorem 3.6 we let $n \to \infty$ and $d \to \infty$ separately, we will have an appropriate intertwining of d and n in concrete situations such that the conditions of the theorem are satisfied. We will demonstrate this procedure explicitly in case of Doppler tomography in Part II of the book.

If $\mathbf{A}$ is not injective then an unavoidable reconstruction error occurs.

Corollary 3.7. *Adopt all assumptions made in Theorem* 3.6. *Then,*

$$\lim_{\substack{n\to\infty \\ d\to\infty}} \|\widetilde{\mathbf{A}}_{n,d}\,\mathbf{A}_n f - f\|_X = \|\mathcal{P}_{\mathsf{N}(\mathbf{A})} f\|_X \,.$$

Proof. The assertion follows from Theorem 3.6 using the estimate

$$\|\widetilde{\mathbf{A}}_{n,d}\,\mathbf{A}_n f - f\|_X \le \|\widetilde{\mathbf{A}}_{n,d}\,\mathbf{A}_n f - \mathcal{P}_{\mathsf{N}(\mathbf{A})^\perp} f\|_X + \|f - \mathcal{P}_{\mathsf{N}(\mathbf{A})^\perp} f\|_X \,,$$

and the fact that $\|f - \mathcal{P}_{\mathsf{N}(\mathbf{A})^\perp} f\|_X = \|\mathcal{P}_{\mathsf{N}(\mathbf{A})} f\|_X$. □

Corollary 3.7 states that $\widetilde{\mathbf{A}}_{n,d}\,\mathbf{A}_n f$ in fact converges to f if $\mathbf{A}$ is injective. The part of f being in the null space $\mathsf{N}(\mathbf{A})$ of $\mathbf{A}$ is invisible for the measurement process and cannot be recovered.

We have demonstrated in Section 2.1 how we obtain reconstruction kernels v_i by means of an inversion formula in case of injective operators. We then even may choose $\epsilon_i = 0$, $i = 1, \ldots, d$. An approximation to v_i according to (3.20) can also be obtained with the help of the SVD, if $\mathbf{A}$ is compact, see Section 2.2. This is important, if $\mathbf{A}$ is not injective or an inversion formula not at hand. If we cut off the series (2.14a) after $M_i < \infty$ steps, then we get an approximation

$$v_{i,M_i} = \sum_{k=0}^{M_i} \sigma_k^{-1} \langle e_i, v_k\rangle_X \, u_k \,, \qquad i = 1, \ldots, d, \tag{3.28}$$

where $\{(\sigma_k, v_k, u_k)\}_{k\in\mathbb{N}_0}$ denotes the singular value decomposition of a compact operator $\mathbf{A} : X \to Y$. We easily see, that

$$\lim_{M_i\to\infty} \|\mathbf{A}^* v_{i,M_i} - \mathcal{P}_{\mathsf{N}(\mathbf{A})^\perp} e_i\|_X = 0 \,,$$

whence (3.20) is satisfied for sufficiently large M_i. We formulate Theorem 3.6 for the special case that v_i is given by (3.28). We omit the proof which can be found in RIEDER, SCHUSTER [101, Theorem 3.12.].

Theorem 3.8. *Let* $\mathbf{A} : X \to Y$ *be compact with SVD* $\{(\sigma_k, v_k, u_k)\}_{k\in\mathbb{N}_0}$. *Further suppose the existence of constants* $0 < \lambda_1 < \lambda_2 < \infty$ *and* $\mu > 0$ *such that*

$$\lambda_1\,(k+1)^{-\mu} < \sigma_k < \lambda_2\,(k+1)^{-\mu} \qquad \text{as } k \to \infty \tag{3.29}$$

and let $\|u_k\|_{Y_1} \leq \kappa\, \sigma_k^{-\beta}$ *for some* $\kappa > 0$, $\beta \geq 0$. *In addition to the requirements from Theorem 3.6 let* $e_i \in \mathsf{D}((\mathbf{A}^*\,\mathbf{A})^{-\alpha}) = \mathsf{R}((\mathbf{A}^*\,\mathbf{A})^{\alpha})$.
If $\alpha > (1+\beta)/2 + 1/(4\,\mu)$ *and* $M_i \geq c\,\rho_n^{-1/(\alpha\,\mu)}$ *for a constant* $c > 0$ *which does not depend on* n, *then there exists a* $C > 0$ *with*

$$\|\widetilde{\mathbf{A}}_{n,d}\,\mathbf{A}_n f - \mathcal{P}_{\mathsf{N}(\mathbf{A})^\perp} f\|_X \leq \|f - E_d f\|_X + C\,\rho_n \left(\frac{1}{d}\sum_{i=1}^{d} \|(\mathbf{A}^*\,\mathbf{A})^{-\alpha} e_i\|_X^2\right)^{1/2} \|f\|_{X_1}\,.$$

In Theorem 3.8 the norms $\|v_i\|_{Y_1}$ are explicitly expressed by

$$\|(\mathbf{A}^*\,\mathbf{A})^{-\alpha} e_i\|_X\,.$$

Note that (3.29) implies a polynomial decrease of the singular values. Hence, severely ill-posed operators $\mathbf{A}$, that means operators with exponentially decreasing singular values, are excluded by the assumptions in Theorem 3.8.

In Theorem 2.6 we investigated how far invariance properties of the underlying operator $\mathbf{A}$ can be used to accelerate the computation of reconstruction kernels and hence to make the whole algorithm more efficient. The questions arises, whether this property is transfered to the condition $\|\mathbf{A}^* v_i - \mathcal{P}_{\mathsf{N}(\mathbf{A})^\perp} e_i\|_X < \epsilon_i$. Is it sufficient to determine only one single kernel which satisfies (3.20) to generate the remaining kernels? The answer is 'yes', if we restrict the invariances to be multiples of an isometry. Moreover, we need an intertwining with respect to $\mathbf{A}^*$ only.

Lemma 3.9. *Let* $\mathbf{A} : X \to Y$, $T : X \to X$ *and* $S : Y \to Y$ *be linear and bounded satisfying* $T\,\mathbf{A}^* = \mathbf{A}^*\,S$. *Further assume that* S *has dense range and that* T *is the multiple of an isometry, that means the existence of a* $\tau > 0$ *with* $\|Tu\|_X = \tau\,\|u\|_X$ *for* $u \in X$. *If* $\|\mathbf{A}^* v - \mathcal{P}_{\mathsf{N}(\mathbf{A})^\perp} e\|_X \leq \epsilon$ *for* $e \in X$, $v \in Y$ *and* $\epsilon > 0$, *then*

$$\|\mathbf{A}^*\,Sv - \mathcal{P}_{\mathsf{N}(\mathbf{A})^\perp}\,Te\|_X \leq \tau\,\epsilon\,.$$

Proof. We only have to show that $\mathcal{P}_{\mathsf{N}(\mathbf{A})^\perp}\,T = T\,\mathcal{P}_{\mathsf{N}(\mathbf{A})^\perp}$. Once this is proved, the assertion follows from

$$\|\mathbf{A}^*\,Sv - \mathcal{P}_{\mathsf{N}(\mathbf{A})^\perp}\,Te\|_X = \|T\,(\mathcal{P}_{\mathsf{N}(\mathbf{A})^\perp} e - \mathbf{A}^* v)\|_X \leq \tau\,\epsilon\,.$$

First we prove the inclusions $T\,\mathsf{N}(\mathbf{A})^\perp \subset \mathsf{N}(\mathbf{A})^\perp$ and $T\,\mathsf{N}(\mathbf{A}) \subset \mathsf{N}(\mathbf{A})$. Let $w \in \mathsf{N}(\mathbf{A})^\perp = \overline{\mathsf{R}(\mathbf{A}^*)}$. Then, there exists a sequence $\{z_k\}$ in Y with $w = \lim_{k\to\infty} \mathbf{A}^* z_k$. Using the invariance property we get $\mathbf{A}^*\,S z_k = T\,\mathbf{A}^* z_k$ and hence $\lim_{k\to\infty} \mathbf{A}^*\,S z_k = Tw$. As a consequence we obtain $Tw \in \overline{\mathsf{R}(\mathbf{A}^*)} = \mathsf{N}(\mathbf{A})^\perp$ which is the first inclusion. Since T/τ is an isometry, we have $T^*\,T = \tau^2\,I_X$. Using that identity and $T\,\mathbf{A}^* = \mathbf{A}^*\,S$ leads to $S^*\,\mathbf{A}\,T = \tau^2\,\mathbf{A}$. Because of $\overline{\mathsf{R}(S)} = Y$, we furthermore have $\mathsf{N}(S^*) = \{0\}$. If $u \in \mathsf{N}(\mathbf{A})$, then by means of all considerations made before we may deduce that $0 = \tau^2\,\mathbf{A}u = S^*\,\mathbf{A}\,Tu$

whence $\mathbf{A}Tu = 0$ follows. This corresponds to the second inclusion. Finally for $x \in X$ we may summarize, that

$$\mathcal{P}_{\mathsf{N}(\mathbf{A})^\perp} Tx = \mathcal{P}_{\mathsf{N}(\mathbf{A})^\perp} T \mathcal{P}_{\mathsf{N}(\mathbf{A})} x + \mathcal{P}_{\mathsf{N}(\mathbf{A})^\perp} T \mathcal{P}_{\mathsf{N}(\mathbf{A})^\perp} x = T \mathcal{P}_{\mathsf{N}(\mathbf{A})^\perp} x ,$$

which completes the proof. □

Up to this moment, we only took measured data $g_n = \mathbf{A}_n f$ into consideration which are free of noise. What happens, if we only have a noise contaminated set of data g_n^η available? In fact, this is a more realistic assumption than to have exact data. Thus, it remains to investigate the stability of the method. To prove that $\widetilde{\mathbf{A}}_{n,d}$ actually is a regularization method in the sense of Definition 1.3, we have to show the existence of a parameter choice rule for d such that the reconstruction error tends to zero with $\eta \to 0$.

We specify the mathematical setup by modelling the noise in the data as a perturbation of the observation operator Ψ_n. This is motivated by the fact that the perturbation of the data are mainly caused by the measurement device. To this end let for $\eta > 0$ the operator $\Psi_n^\eta : Y_1 \to \mathbb{K}^n$ be defined as

$$(\Psi_n^\eta w)_k = (\Psi_n w)_k + \eta_k \, \|w\|_{Y_1} , \quad \eta_k \le \eta , \quad k = 1, \dots, n , \tag{3.30}$$

what implies $n^{-1/2} \|(\Psi_n^\eta - \Psi_n) w\|_2 / \|w\|_{Y_1} \le \eta$. We outline that $\widetilde{\mathbf{A}}_{n,d}$ has a regularizing effect using an appropriate coupling of d and the number of data n. Note that setting (3.30) yields that the relative noise level is bounded by η, but we did not specify the particular kind of noise.

Theorem 3.10. *Beyond the assumptions made in Theorem 3.6 we require that the triples*

$$\{(e_i, v_i, b_i)\}_{i=1}^d \subset X \times Y_1 \times X$$

allow for a coupling of d and n such that $d = d_n \to \infty$ for $n \to \infty$ and

$$\lim_{n\to\infty} \rho_n^2 \, d_n^{-1} \sum_{i=1}^{d_n} \|v_i\|_{Y_1}^2 = 0$$

is valid as well as

$$\lim_{n\to\infty} d_n^{-1} \sum_{i=1}^{d_n} \epsilon_i^2 = 0 .$$

If furthermore $n = n_\eta$ in such a way that $n_\eta \to \infty$ and $\eta / \rho_{n_\eta} = O(1)$ for $\eta \to 0$, then

$$\limsup_{\eta\to 0} \left\{ \|\widetilde{\mathbf{A}}_{n_\eta, d_{n_\eta}} w - \mathcal{P}_{\mathsf{N}(\mathbf{A})^\perp} f\|_X : w = \Psi_{n_\eta}^\eta \, \mathbf{A} f , \ \Psi_{n_\eta}^\eta \ \text{satisfies (3.30)} \right\} = 0$$

for all $f \in X_1$.

Proof. We denote by $g_n = \Psi_n \mathbf{A} f$ a set of exact data and by $g_n^\eta = \Psi_n^\eta \mathbf{A} f$ a set of noisy data according to (3.30). Since the family $\{b_i\}$ is a Riesz system we can use the second inequality in (3.6) to get

$$\|\widetilde{\mathbf{A}}_{n,d}(g_n - g_n^\eta)\|_X \le C\, d^{-1/2} \Big(\sum_{i=1}^d |\langle (\Psi_n - \Psi_n^\eta)\, \mathbf{A} f, G_n \Psi_n v_i \rangle_{\mathbb{K}^n}|^2 \Big)^{1/2}$$
$$= C\, d^{-1/2} \Big(\sum_{i=1}^d |\langle G_n^{1/2} (\Psi_n - \Psi_n^\eta)\, \mathbf{A} f, G_n^{1/2} \Psi_n v_i \rangle_{\mathbb{K}^n}|^2 \Big)^{1/2}.$$

Since $\{\phi_k\}$ is a Riesz system too, we can estimate the spectral norm of the Gramian matrix G_n by means of (3.6) as

$$\|G_n\|_2 \le \kappa / n$$

for some constant $\kappa > 0$.

Together with (3.24), (3.25), (3.30) and the continuity of $\mathbf{A} : X_1 \to Y_1$ both estimates lead us to

$$\|\widetilde{\mathbf{A}}_{n,d}(g_n - g_n^\eta)\|_X \le C\, d^{-1/2}\, \|(\Psi_n - \Psi_n^\eta)\, \mathbf{A} f\|_{\mathbb{K}^n}\, \|G_n^{1/2}\|_2$$
$$\times \Big(\sum_{i=1}^d \|G_n^{1/2} \Psi_n v_i\|_{\mathbb{K}^n}^2 \Big)^{1/2}$$
$$\le C\, d^{-1/2}\, \eta\, \|\mathbf{A} f\|_{Y_1} \sqrt{\kappa/n} \Big(\sum_{i=1}^d c^{-1}\, n\, \|\Pi_n v_i\|_Y^2 \Big)^{1/2}$$
$$\le \frac{\sqrt{\kappa}\, C\, C_b}{\sqrt{c}}\, \eta\, \|\mathbf{A}\|_{X_1 \to Y_1} \|f\|_{X_1} \Big(\frac{1}{d} \sum_{i=1}^d \|v_i\|_{Y_1}^2 \Big)^{1/2}.$$

Using the convergence estimate (3.27) we finally obtain

$$\|\widetilde{\mathbf{A}}_{n,d} g_n^\eta - \mathcal{P}_{\mathsf{N}(\mathbf{A})^\perp} f\|_X \le \|\widetilde{\mathbf{A}}_{n,d}(g_n^\eta - g_n)\|_X + \|\widetilde{\mathbf{A}}_{n,d} g_n - \mathcal{P}_{\mathsf{N}(\mathbf{A})^\perp} f\|_X$$
$$\le \|(I - E_d)\, \mathcal{P}_{\mathsf{N}(\mathbf{A})^\perp} f\|_X + \tilde{c} \Big[(\eta + \rho_n) \Big(\frac{1}{d} \sum_{i=1}^d \|v_i\|_{Y_1}^2 \Big)^{1/2} + \Big(\frac{1}{d} \sum_{i=1}^d \epsilon_i^2 \Big)^{1/2} \Big] \|f\|_{X_1}$$

with a suitable constant $\tilde{c} > 0$. The proof is finished replacing n by n_η and d by d_{n_η}. □

In Part II of the book, we consider the problem of vector field tomography as a specific application and demonstrate the existence of sequences n_η and d_{n_η} satisfying the assumptions of Theorem 3.10.

4

Approximate inverse in distribution spaces

There exist inverse problems which can no longer be represented by an operator equation neither on Hilbert nor on Banach spaces. This situation appears, if an integral operator does not assign smooth functions to smooth functions again or if the elements contained in the range are not integrable. An example for such an operator is given by the *spherical Radon transform*, when the center set is a hyperplane. In contrast to the classical Radon transform, the spherical Radon transform – which is also known as the *spherical mean operator* – maps a function to its integrals over spheres. It serves as a mathematical model for problems in SAR and SONAR and hence is of great practical relevance. Even if the function f is rapidly decreasing, the spherical means $\mathbf{M}f$ are not even integrable. The mathematical properties of $\mathbf{M}$ are outlined in detail in Part III.

Thus, there is motivation to investigate equations

$$\mathbf{A} : V' \to W'$$

and their regularizations thoroughly, where V' and W' are dual spaces of certain function spaces. Concrete regularization methods for mappings between distribution spaces are barely found in literature. This chapter aims to extend the method of approximate inverse to such mappings. To do so, we have to give a new definition of what we mean by a *mollifier* in a distributional sense and this definition plays a key role in our investigations. We will state such a definition which turns out to be a weakening compared to its introduction in Definition 2.1. We show that the method of approximate inverse has all advantages we know from the Hilbert space setting: The reconstruction kernels can be precomputed independently of the measurement process and hence are not affected by any noise in the measured data, and invariances of the underlying operator $\mathbf{A}$ can be used to increase the efficiency of the method.

Certainly, we would like to transfer the convergence and stability analysis made in Chapter 3 to the distributional case. But then we run into the difficulty that the important term of a Riesz basis cannot be translated to distribution spaces meaningfully. Nevertheless, we will give a brief sketch on how to deal with semi-discrete problems.

In this chapter we do not adapt the theory to Sobolev spaces of negative order which also contain distributions. The reason is that there are integral operators, as e.g. **M**, which can not be formulated as continuous mappings between Sobolev spaces of negative order.

4.1 Mollifier and reconstruction kernels in dual spaces of smooth functions

At first, we specify the function spaces V and W. Let $\Omega_1 \subset \mathbb{K}^{n_1}$ und $\Omega_2 \subset \mathbb{K}^{n_2}$ be open domains and $\mathbb{K}$ the field of real or complex numbers. Assume $V \subset \mathcal{C}^\infty(\Omega_1)$, $W \subset \mathcal{C}^\infty(\Omega_2)$ to be subspaces of smooth functions, which are closed with respect to their topology. As examples, one can see the space of rapidly decreasing functions $\mathcal{S}(\mathbb{R}^m)$ or the *Schwartz space* $\mathcal{D}(\mathbb{R}^m) = \mathcal{C}_0^\infty(\mathbb{R}^m)$. Further suppose that $\mathbf{A} : V' \to W'$ is linear, continuous and injective. With V', W' we denote the dual spaces associated to V, W, that is the spaces consisting of all linear functionals $V \to \mathbb{K}$, $W \to \mathbb{K}$ which are continuous in the topology of V and W, respectively. We consider the inverse problem to find a distribution $f \in V'$ which for given $g \in W'$ fulfills

$$\mathbf{A}f = g\,. \tag{4.1}$$

If we want to construct a regularization method for (4.1), then we have to take into account that there do not exist any inner products and orthogonal projections on V', W'. But the orthogonal projection plays a crucial part when defining regularization schemes in Hilbert spaces. Since a generalized inverse of $\mathbf{A}$ is not defined, it is not entirely clear what we mean by a *regularization method* for operators between distribution spaces. Nevertheless, the introduction of *mollifiers* makes perfect sense also in that case. We denote by V'', W'' the double duals of V, W; these are the duals of V', W' if we endow the latter ones with the weak $*$-topology.

Definition 4.1. *Let $\gamma > 0$. Assume $e_\gamma(y) \in V''$ for all $y \in \Omega_1$ be given, such that*

$$\lambda_\gamma(y) := \langle \lambda, e_\gamma(y) \rangle_{V' \times V''} \in V' \qquad \text{for all } \lambda \in V' \tag{4.2}$$

holds true. We call e_γ a mollifier, *if in addition to (4.2) the convergence*

$$\lim_{\gamma \to 0} \langle \lambda_\gamma, \beta \rangle_{V' \times V} = \langle \lambda, \beta \rangle_{V' \times V} \qquad \text{for all } \beta \in V \tag{4.3}$$

is valid. For subspaces $V_1 \subset V'$ and $V_2 \subset V$ we call e_γ a (V_1, V_2)-mollifier, *iff (4.2) is satisfied for all $\lambda \in V_1$ and condition (4.3) holds true for all $\lambda \in V_1$ and $\beta \in V_2$.*

As in Definition 4.1 we always denote dual pairings with $\langle \cdot, \cdot \rangle_{V' \times V}$ and $\langle \cdot, \cdot \rangle_{V' \times V''}$.

Hence, if e_γ is a mollifier in the sense of Definition 4.1 and $f \in V'$ a solution of (4.1), then

$$f_\gamma(y) := \langle f, e_\gamma(y) \rangle_{V' \times V''}, \quad y \in \Omega_1$$

is a distribution in V', which converges to f with respect to the weak $*$-topology given on V'. Since $V \subset V''$ also with respect to the topology (see Rudin [105, Section 4.5]), we may choose e_γ in V yielding a smooth approximation f_γ for f. For this reason, Definition 4.1 is a meaningful and straightforward extension of the concept of mollifers to distributions.

In accordance with the approximate inverse in L^2-spaces the approximations f_γ will be computed with the help of *reconstruction kernels* also in the distributional case. The operator $\mathbf{A}$ has a linear, continuous adjoint $\mathbf{A}^* : W'' \to V''$ defined by

$$\langle \mathbf{A}^* w, v \rangle_{V'' \times V'} = \langle w, \mathbf{A} v \rangle_{W'' \times W'}, \qquad w \in W'', \quad v \in V',$$

which has a dense range $\mathsf{R}(\mathbf{A}^*)$ in V'' because of the injectivity of $\mathbf{A}$. Assuming $e_\gamma(y) \in \mathsf{R}(\mathbf{A}^*)$ for all $y \in \Omega_1$, then the equation

$$\mathbf{A}^* v_\gamma(y) = e_\gamma(y), \qquad y \in \Omega_1 \tag{4.4}$$

has a solution $v_\gamma(y)$ and $f_\gamma(y)$ computes as

$$f_\gamma(y) = \langle f, \mathbf{A}^* v_\gamma(y) \rangle_{V' \times V''} = \langle g, v_\gamma(y) \rangle_{W' \times W''}, \qquad y \in \Omega_1 .$$

This motivates the following extension of the method of approximate inverse.

Definition 4.2. *Assume* $\mathbf{A} : V' \to W'$ *to be linear, continuous and injective and* e_γ *to be a mollifier according to Definition* 4.1. *Furthermore let* $e_\gamma(y) \in \mathsf{R}(\mathbf{A}^*)$ *for all* $y \in \Omega_1$. *The mapping* $\widetilde{\mathbf{A}}_\gamma : W' \to V'$ *defined by*

$$\widetilde{\mathbf{A}}_\gamma w(y) = \langle w, v_\gamma(y) \rangle_{W' \times W''}, \qquad w \in W', \; y \in \Omega_1, \tag{4.5}$$

where $v_\gamma(y)$ *solves* (4.4), *is called* (distributional) approximate inverse *of* $\mathbf{A}$.

Note that Definition 4.1 implies that $\widetilde{\mathbf{A}}_\gamma$ is well-defined: For $w \in W'$ we always have $\widetilde{\mathbf{A}}_\gamma w \in V'$.

Remark 4.3. The particular choice of $e_\gamma(y) \subset V$ in general does not automatically imply $v_\gamma(y) \in W$ and hence the reconstruction kernel being smooth (if it exists at all). Thus, the question arises: When does (4.4) have a solution contained in W, supposed that the mollifier is in V? This is likely – but not guaranteed – if $\mathsf{R}(\mathbf{A}^*) \cap V$ is dense in V. This is the case, if

a) the function spaces V and W are reflexive, that means we have $V = V''$, $W = W''$, also with respect to the topology. In this situation $\mathbf{A}^* : W \to V$ is linear, continuous with dense range.
b) we have $\mathbf{A}^*(W) \subset V \subset V''$.

We now show that the name *approximate inverse* actually is justified for $\widetilde{\mathbf{A}}_\gamma$.

Lemma 4.4. *If $g \in \mathsf{R}(\mathbf{A})$, $\mathbf{A} \in \mathcal{L}(V', W')$ is one-to-one and $e_\gamma(y) \in \mathsf{R}(\mathbf{A}^*)$ for all $y \in \Omega_1$, then*

$$\lim_{\gamma \to 0} \widetilde{\mathbf{A}}_\gamma g = f \qquad \text{in } V',$$

where $\mathbf{A} f = g$.

Proof. Since $\mathbf{A}$ is one-to-one, $g \in \mathsf{R}(\mathbf{A})$, there is a unique f with $\mathbf{A} f = g$. The condition for e_γ assures that (4.4) is solvable. The convergence finally follows from (4.5) and the fact that e_γ is a mollifier in the sense of Definition 4.1. □

Condition $e_\gamma(y) \in \mathsf{R}(\mathbf{A}^*)$, $y \in \Omega_1$, seems to be rather restrictive at first glance. However, in applications equation (4.4) is to be solved for a finite number of reconstruction points $y_i \in \Omega_1$ only. Furthermore we are eager again to use invariances of $\mathbf{A}$ so that (4.4) possibly has to be solved only once. To do so we have to take into account that in the distributional case we rely upon invariances for the adjoint $\mathbf{A}^*$ since a normal equation is no longer available.

Lemma 4.5. *Assume that $T_1^y \in \mathcal{L}(V'')$, $T_2^y \in \mathcal{L}(W'')$ are linear and continuous for $y \in \Omega_1$ and that there exists a $y^\star \in \Omega_1$ with $e_\gamma(y) = T_1^y e_\gamma(y^\star)$, where $e_\gamma(y^\star) \in \mathsf{R}(\mathbf{A}^*)$. If*

$$T_1^y \mathbf{A}^* = \mathbf{A}^* T_2^y, \qquad y \in \Omega_1, \tag{4.6}$$

then

$$\upsilon_\gamma(y) = T_2^y \upsilon_\gamma(y^\star)$$

are reconstruction kernels associated with $e_\gamma(y)$ whenever $\upsilon_\gamma(y^\star)$ satisfies

$$\mathbf{A}^* \upsilon_\gamma(y^\star) = e_\gamma(y^\star).$$

Proof. Because of $e_\gamma(y^\star) \in \mathsf{R}(\mathbf{A}^*)$ there is a $w_\gamma \in W''$ with $\mathbf{A}^* w_\gamma = e_\gamma(y^\star)$. Applying (4.6), we obtain

$$e_\gamma(y) = T_1^y e_\gamma(y^\star) = T_1^y \mathbf{A}^* w_\gamma = \mathbf{A}^* T_2^y w_\gamma,$$

whence $e_\gamma(y) \in \mathsf{R}(\mathbf{A}^*)$ for $y \in \Omega_1$. Hence, equation (4.4) is solvable and using (4.6) again we may deduce that

$$e_\gamma(y) = T_1^y e_\gamma(y^\star) = T_1^y \mathbf{A}^* \upsilon_\gamma(y^\star) = \mathbf{A}^* T_2^y \upsilon_\gamma(y^\star).$$

□

Remark 4.6. We summarize that the distributional approximate inverse (4.5) has all features which we know from the Hilbert spaces setting: The reconstruction kernels may be precomputed as solutions of (4.4) and hence are independent from the noise level in the measurement data. Furthermore invariance properties can be used to improve the efficiency of the algorithm. If we had no such invariances for the spherical Radon transform, then an application of this method as inversion scheme for $\mathbf{M}$ would not pay because of the large computation time.

It is still an open question what to do when $e_\gamma(y) \notin \mathsf{R}(\mathbf{A}^*)$. A normal equation does not exist and would anyway be equivalent to (4.4) in case that $\mathbf{A}$ is injective. SCHÖPFER ET AL. [107] formulated a Landweber-type method to solve operator equations in Banach spaces, but it is unclear how this method extends to distribution spaces. Since $\mathbf{A}^*$ has dense range in V'', we find to a finite number of reconstruction points $y_i \in \Omega_1$, $i = 1, \dots, d$, and given bounds $\epsilon_i > 0$, $i = 1, \dots, d$, as well as elements $v_\gamma^i \in W''$ satisfying

$$|\langle \mathbf{A}^* v_\gamma^i - e_\gamma(y_i), \lambda \rangle_{V'' \times V'}| < \epsilon_i \qquad \text{for all } \lambda \in V' , \quad i = 1, \dots, d . \tag{4.7}$$

This is an equivalent formulation of (3.20) in the weak $*$-topology of V''. By now, we do not know how condition (4.7) might be checked in a concrete application.

We complete this section by comparing the concept of a mollifier given in Definition 4.1 with that for L^2-spaces as stated in Definition 2.1. If $V = L^2(\Omega_1)$, then $V' = V'' = V$, since $L^2(\Omega_1)$ is a Hilbert space. Applying Definition (4.1) to V we obtain the weak L^2-convergence

$$\lim_{\gamma \to 0} \langle f_\gamma, v \rangle_{L^2(\Omega_1)} = \langle f, v \rangle_{L^2(\Omega_1)} \qquad \text{for all } v \in L^2(\Omega_1) ,$$

as condition for e_γ to be a mollifier, where $f_\gamma(y) = \langle f, e_\gamma(\cdot, y) \rangle_{L^2(\Omega_1)}$. From this view Definition 4.1 includes a weakening of the mollifier concept compared to 2.1 which is due to the weak $*$-topology defined on V'. This emphasizes that the idea of a mollifier is always bound to a corresponding topology.

Remark 4.7. Even though we introduced V and W as subspaces of smooth functions, all the considerations made in this section can be transfered to other pairings of function spaces and their duals without any difficulties. For instance such pairings are given by $V = \mathcal{C}(K)$, where $K \subset \mathbb{R}^{n_1}$ is compact, with its dual space V' consisting of the regular Borel measures on K, or the pairing $V = L^p(\Omega_1)$, $1 \le p < \infty$, with $V' = L^q(\Omega_1)$, $p^{-1} + q^{-1} = 1$, or Sobolev spaces $V = H^\alpha(\Omega_1)$ and $V' = H^{-\alpha}(\Omega_1)$. In all these cases the definitions 4.1 and 4.2 lead to an approximate inverse in the sense of Lemma 4.4, where the convergence always is to be understood with respect to the weak topology.

4.2 Dealing with semi-discrete equations

This section is dedicated to establishing a semi-discrete setting similar to that in Chapter 3. To this end we again introduce an *observation operator* $\Psi_n : W' \to \mathbb{K}^n$ representing the measurement process. Let n linear, bounded functionals $\psi_{n,k} \in W''$, $k = 1, \ldots, n$ be given such that

$$(\Psi_n w)_k = \langle \psi_{n,k}, w \rangle_{W'' \times W'}\,, \qquad k = 1, \ldots, n\,. \tag{4.8}$$

We seek a distribution $f \in V'$ satisfying

$$\mathbf{A}_n f = g_n\,, \tag{4.9}$$

where $\mathbf{A}_n = \Psi_n \, \mathbf{A}$ and $g_n = \Psi_n g \in \mathbb{K}^n$ is the given outcome of the measurement process.

If we tried to carry over the concepts for solving semi-discrete operator equations in Hilbert spaces outlined in Chapter 3 to distribution spaces, we would have the problem that the fundamental concept of Riesz systems is not available in those spaces. Nevertheless, we try to establish a theory of dealing with equations like (4.9) in distribution spaces. The following investigations are first steps in this direction.

We start by defining a mollifier operator E_d as we did in (3.5). Let $d \in \mathbb{N}$ and $\{e_i\}_{i=1}^d \subset V''$, $\{\mathrm{v}_i\}_{i=1}^d \subset V'$ be sequences. The mapping $E_d : V' \to V'$

$$E_d f = \sum_{i=1}^{d} \langle f, e_i \rangle_{V' \times V''} \, \mathrm{v}_i \tag{4.10}$$

has the *mollifier property*, if

$$\lim_{d \to \infty} E_d f = f \qquad \text{in } V' \quad \text{for all } f \in V'\,. \tag{4.11}$$

The limit (4.11) is to be understood with respect to the weak $*$-topology in V' induced from V, that is

$$\lim_{d \to \infty} \langle E_d f, \beta \rangle_{V' \times V} = \langle f, \beta \rangle_{V' \times V} \qquad \text{for all } f \in V'\,, \quad \beta \in V\,.$$

Furthermore we assume that $e_i \in \mathsf{R}(\mathbf{A}^*)$, $i = 1, \ldots, d$ and that hence the equations

$$\mathbf{A}^* v_i = e_i\,, \qquad i = 1, \ldots, d \tag{4.12}$$

have solutions in W''.

In accordance to Section 3.1, we postulate a coupling of the observation operator Ψ_n to an interpolation operator $\Pi_n : W' \to W'$ which has to satisfy two requirements. More precisely,

$$\Pi_n w = \sum_{k=1}^{n} \langle \psi_{n,k}, w \rangle_{W'' \times W'} \, \mathrm{w}_k\,,$$

where the sequence $\{\mathrm{w}_k\}_{k=1}^n \subset W'$ is assumed to be defined for arbitrary $n \in \mathbb{N}$. On the one side, the interpolation Π_n is required to obey a boundedness condition

$$|\langle \Pi_n w, z\rangle_{W' \times W''}| \le C_b \, |\langle w, z\rangle_{W' \times W''}| \qquad \text{for all } w \in W', \quad z \in W'' \quad (4.13)$$

with a constant $C_b > 0$. On the other side, we postulate the convergence

$$\lim_{n\to\infty} |\langle w - \Pi_n w, z\rangle_{W' \times W''}| = 0 \qquad \text{for all } w \in W', \quad z \in W''. \quad (4.14)$$

In other words: We have that $I - \Pi_n \to 0$ for $n \to \infty$ in W' with respect to the weak topology induced from W''.

The aim is to gain reconstruction kernels $v_i^n \in \mathbb{K}^n$ with the help of solutions v_i of (4.12) and then to define a semi-discrete approximate inverse. In Section 3.1 we have done this by means of Ψ_n. That was possible since Hilbert spaces are reflexive and hence the kernels v_i were in Y. The mappings Ψ_n (4.8), however, are defined on W' and hence are not suited to generate the discrete kernels v_i^n. Therefore, we introduce a further operator $\Pi_n' : W'' \to W''$. Assume we have functionals $\psi_{n,k}'$ in W'''[1] as well as some elements $\mathrm{w}_k' \in W''$, $k = 1, \ldots, n$, $n \in \mathbb{N}$, such that

$$\Pi_n' w = \sum_{k=1}^{n} \langle \psi_{n,k}', w\rangle_{W''' \times W''} \, \mathrm{w}_k' \,, \qquad w \in W''$$

satisfies the convergence

$$\lim_{n\to\infty} |\langle w, z - \Pi_n' z\rangle_{W' \times W''}| = 0 \qquad \text{for all } w \in W', \quad z \in W'', \quad (4.15)$$

that means, $I - \Pi_n' \to 0$ for $n \to \infty$ in the weak $*$-topology of W''.

Remark 4.8. We virtually have free choice for the functionals $\psi_{n,k}' : W'' \to \mathbb{K}$. It is only important that there are sequences $\{\mathrm{w}_k'\}_{k=1}^n \in W''$ for $n \in \mathbb{N}$ satisfying the convergence (4.15). A possible choice for $\{\psi_{n,k}'\}_{k=1}^n$ is the dual basis of $\{\psi_{n,k}\}_{k=1}^n$.

Finally, we denote $\Psi_n' : W'' \to \mathbb{K}^n$ by

$$(\Psi_n' w)_k = \langle \psi_{n,k}', w\rangle_{W''' \times W''} \,, \qquad w \in W''.$$

We get the kernels $v_i^n \in \mathbb{K}^n$ for the semi-discrete problem (4.9) setting

$$v_i^n := G_n \, \Psi_n' v_i \,, \qquad i = 1, \ldots, d.$$

Here, $G_n \in \mathbb{K}^{n \times n}$ is the Gramian matrix with respect to the families $\{\mathrm{w}_k\}$, $\{\mathrm{w}_k'\}$,

$$(G_n)_{k,l} = \langle \mathrm{w}_k, \mathrm{w}_l'\rangle_{W' \times W''} \,, \qquad 1 \le k, l \le n.$$

[1] With W''' we denote the dual space of W'' with respect to the weak $*$-topology given in W''.

Theorem 4.9. *Let $\mathbf{A} : V' \to W'$ be a linear, bounded and one-to-one operator and let $\mathbf{A}_n = \Psi_n \mathbf{A}$ for a linear and continuous observation operator $\Psi_n : W' \to \mathbb{K}^n$. We further assume the existence of triples*

$$\{(e_i, v_i, \mathrm{v}_i)\}_{i=1}^d \subset V'' \times W'' \times V' \quad und \quad \{(\mathrm{w}_k, \psi'_{n,k}, \mathrm{w}'_k)\}_{k=1}^n \subset W' \times W''' \times W''$$

such that the mappings E_d, Π_n and Π'_n satisfy the conditions (4.11), (4.12), (4.13), (4.14) *and* (4.15). *Then the* (distributional) semi-discrete approximate inverse $\widetilde{\mathbf{A}}_{n,d} : \mathbb{K}^n \to V'$ *given by*

$$\widetilde{\mathbf{A}}_{n,d} v = \sum_{i=1}^d \langle v, G_n \Psi'_n v_i \rangle_{\mathbb{K}^n} \mathrm{v}_i$$

has the convergence property

$$\lim_{d\to\infty} \lim_{n\to\infty} \widetilde{\mathbf{A}}_{n,d} \mathbf{A}_n f = f \qquad in\ V' \tag{4.16}$$

for all $f \in V'$.

Proof. Let $\beta \in V$. We may estimate

$$|\langle \widetilde{\mathbf{A}}_{n,d} \mathbf{A}_n f - f, \beta \rangle_{V' \times V}| \le |\langle E_d f - f, \beta \rangle_{V' \times V}| + |\langle \widetilde{\mathbf{A}}_{n,d} \mathbf{A}_n f - E_d f, \beta \rangle_{V' \times V}| .$$

Note that the first part on the right-hand side does not depend on n. Taking into consideration (4.11), it is sufficient to prove

$$\lim_{n\to\infty} |\langle \widetilde{\mathbf{A}}_{n,d} \mathbf{A}_n f - E_d f, \beta \rangle_{V' \times V}| = 0$$

for arbitrary, fixed $d \in \mathbb{N}$. Since

$$\begin{aligned} &|\langle \widetilde{\mathbf{A}}_{n,d} \mathbf{A}_n f - E_d f, \beta \rangle_{V' \times V}| \\ &\qquad \le \sum_{i=1}^d \left\{ |\langle f, e_i \rangle_{V' \times V''} - \langle \mathbf{A}_n f, G_n \Psi'_n v_i \rangle_{\mathbb{K}^n}| \, |\langle \mathrm{v}_i, \beta \rangle_{V' \times V}| \right\} \end{aligned} \tag{4.17}$$

we have to show, that

$$\lim_{n\to\infty} \left\{ |\langle f, e_i \rangle_{V' \times V''} - \langle \mathbf{A}_n f, G_n \Psi'_n v_i \rangle_{\mathbb{K}^n}| \right\} = 0 .$$

First, we find that

$$\langle \mathbf{A}_n f, G_n \Psi'_n v_i \rangle_{\mathbb{K}^n} = \langle \Pi_n \mathbf{A} f, \Pi'_n v_i \rangle_{W' \to W''}$$

is valid. Equation (4.12) yields $\langle f, e_i \rangle_{V' \times V''} = \langle \mathbf{A} f, v_i \rangle_{W' \times W''}$ which leads together with (4.13) to

$$\begin{aligned} &|\langle \mathbf{A} f, v_i \rangle_{W' \times W''} - \langle \Pi_n \mathbf{A} f, \Pi'_n v_i \rangle_{W' \times W''}| \\ &\qquad \le |\langle (I - \Pi_n) \mathbf{A} f, v_i \rangle_{W' \times W''}| + |\langle \Pi_n \mathbf{A} f, (I - \Pi'_n) v_i \rangle_{W' \times W''}| \\ &\qquad \le |\langle (I - \Pi_n) \mathbf{A} f, v_i \rangle_{W' \times W''}| + C_b \, |\langle \mathbf{A} f, (I - \Pi'_n) v_i \rangle_{W' \times W''}| . \end{aligned}$$

The assertion can then be deduced from the convergence properties (4.14) and (4.15) of Π_n and Π'_n, respectively. □

Remark 4.10. The order of limits in (4.16) is essential and must not be changed. This can be seen from (4.17) and is an important difference compared to the convergence proof of Theorem 3.6. This is the price we have to pay because there is no analogue to the Riesz property for the systems $\{\mathrm{v}_i\}$, $\{\mathrm{w}_k\}$ or $\{\mathrm{w}'_k\}$.

To finish this section, we consider a particular situation which is of great importance for the spherical Radon transform **M** and easily allows for solving semi-discrete problems.

Let $\mathsf{R}(\mathbf{A}) \subset (\mathcal{C}(\Omega_2) \cap W')$ and the observation operator Ψ_n be defined via point evaluations

$$(\Psi_n w)_k = \langle \psi_{n,k}, w \rangle_{W'' \times W'} = w(\theta_k)\,, \qquad w \in \mathsf{R}(\mathbf{A})\,, \quad k = 1, \ldots, n\,,$$

where $\theta_k \in \Omega_2$, $k = 1, \ldots, n$, are given scanning points. Suppose further that **A** satisfies an invariance property as in Lemma 4.5. We choose a mollifier $e_\gamma(y^\star) \in \mathsf{R}(\mathbf{A}^*)$ and assume that the solution $v_\gamma(y^\star)$ of $\mathbf{A}^* v_\gamma(y^\star) = e_\gamma(y^\star)$ belongs to $\mathcal{C}(\Omega_2) \cap W''$, hence is a continuous function. For arbitrary $y \in \Omega_1$ we have $v_\gamma(y) = T_2^y v_\gamma(y^\star)$ for a certain $T_2^y \in \mathcal{L}(W'')$. The semi-discrete approximate inverse to solve

$$\Psi_n \, \mathbf{A} f = \mathbf{A}_n f = g_n$$

for given $g_n \in \mathbb{K}^n$ can then be formulated as

$$\widetilde{\mathbf{A}}_{n,\gamma} g_n(y) = \langle g_n, Q_n \, \Psi_n \, T_2^y v_\gamma(y^\star) \rangle_{\mathbb{K}^n} \tag{4.18}$$

and emerges from the continuous approximate inverse

$$\widetilde{\mathbf{A}}_\gamma g(y) = \langle g, T_2^y v_\gamma(y^\star) \rangle_{W' \times W''}\,, \qquad g \in \mathsf{R}(\mathbf{A})\,, \tag{4.19}$$

by applying a numerical integration rule, as for example the trapezoidal sum corresponding to the nodes $\{\theta_k\}$. This is possible, since the dual pairing on the right-hand side of (4.19) is the L^2-inner product and the convex hull of the scanning points $\{\theta_k\}$ is a compact and hence bounded subset of $\mathbb{K}^{n_2}$. The matrix entries of $Q_n \in \mathbb{K}^{n \times n}$ are the weights of the applied integration rule.

5

Conclusion and perspectives

In this first part of the book, we have introduced the method of approximate inverse and have shown how it can be used to solve semi-discrete operator equations in various situations. We have further emphasized the different features of the method. Since we have complete freedom in choosing the mollifier, the approximate inverse represents a flexible tool to solve inverse problems. Since the reconstruction kernels are computed independently of the measurement process, this method is well suited for large-scale computations.

The mathematical framework of the approximate inverse might offer a possibility to develop a unified theory for regularization methods. The idea is to assign each regularization method a mollifer such that the resulting approximate inverse and the given regularization coincide. A big step in that direction can be seen in the article LOUIS [68]. There, filter methods and the approximate inverse are identified as smoothing the generalized inverse $\mathbf{A}^\dagger$ on the one hand and as applying the generalized inverse to smoothed data on the other hand.

A rigorous extension of the convergence and stability analysis of the approximate inverse for mappings between distribution spaces will further be subject of future research. First ideas have been presented in Section 4.2. The application of the method to the spherical means operator is outlined in Part III. First numerical results are promising.

Part II

Application to 3D Doppler Tomography

Vector and tensor tomography is a relatively new kind of tomography compared to conventional X-ray CT. Vector field tomography is the reconstruction of the velocity or vorticity of a moving fluid or gas from a set of integrals. These integrals are obtained e.g. by time-of-flight measurements or applying an electric field. When the measurement procedure uses ultrasound and relies on the *Doppler effect*, then we call this particular vector tomography problem *Doppler tomography.* Applications of vector and tensor tomography include medical diagnosis, oceanography, plasma physics or photoelasticity to name only a few. We refer to SPARR AND STRÅHLÉN [119] for a comprehensive overview.

The measured data often are integrals along lines over projections of the vector field:

$$y(\mathbf{f}, L) = \int_L \langle \theta, \mathbf{f}(x) \rangle \, d\ell(x) \, .$$

Here θ is a unit vector. In Doppler tomography $\theta = \theta_L$ is the vector of direction corresponding to the line L. The mapping $\mathbf{f} \mapsto y(\mathbf{f}, \cdot) =: \mathbf{Df}$ is then called *Doppler transform.* It is equal to the first moment of the velocity spectrum and coincides with the longitudinal ray transform for vector fields in SHARAFUTDINOV [116]. JUHLIN [50] outlined how to get the Doppler transform of a searched for velocity field $\mathbf{f}$ as data with the help of ultrasound Doppler measurements. In JANSSON ET AL. [47] the authors developed an experimental setup at the Lund Institute of Technology (Sweden). The inverse problem thus consists of recovering the flow $\mathbf{f}$ from its Doppler transform. In SPARR ET AL. [120], SHARAFUTDINOV [116] and [109] the authors investigate the mathematical properties of $\mathbf{D}$. As an important result they state that only the *solenoidal part* of $\mathbf{f}$ can be detected from $\mathbf{Df}$ since the Doppler transform has a non-trivial null space consisting of potential fields and hence is not injective. This is an essential difference from the Radon transform.

In recent years various methods to recover vector and tensor fields from integral measurements have been established. In NORTON [88], BRAUN AND HAUCK [12] and SHARAFUTDINOV [116], the authors prove inversion formulas which yield the solenoidal part of $\mathbf{f}$. WINTERS, ROUSEFF [131] present an algorithm of filtered backprojection type to compute the curl of $\mathbf{f}$. In OSMAN, PRINCE [90] a vector field tomography problem in three dimensions on bounded domains with arbitrary boundary conditions is considered. DESBAT [24] investigates optimal sampling schemes by transferring ideas from scalar computerized tomography. An iterative scheme for inverting $\mathbf{D}$ based on algebraic reconstruction techniques is outlined in WERNSDÖRFER [129]. Least squares methods are considered in BEZUGLOVA ET AL. [9] and DEREVTSOV AND KASHINA [22]. ANDERSSON [4] uses higher moments of the velocity spectrum to reconstruct not only the solenoidal part but also irrotational parts of $\mathbf{f}$. In [109, 110] the author applies the concept of approximate inverse to the Doppler tomography problem yielding a method of filtered backprojection type.

As mentioned before, the applications of vector tomography problems are not confined to medicine. SIELSCHOTT [118] investigates the gas dynamics of a furnace using time-of-flight-measurements. STEFANI, GERBETH [121, 122] present a mathematical model to reconstruct velocities of electroconductive melts by measuring magnetic fields and electrical potentials which are induced by a disposed electric field. They deal with an integral equation which arises from Maxwell's equations and the Biot-Savart law and is not related to the Doppler transform.

The intention of this part of the book is to apply the concepts established in part 1 to the three-dimensional Doppler transform. In the first chapter we briefly outline how to get the Doppler transform of a velocity field by sending ultrasound waves. After that we introduce the semi-discrete Doppler transform corresponding to the fact that we have only a finite number of data available. Hence, the semi-discrete setup is relevant from the practical point of view. In the next chapter we provide all ingredients being necessary to apply the method of semi-discrete approximate inverse to the Doppler tomography problem. These include the mollifier operator $\mathbf{E}_d$ which is important for the convergence of the method and the interpolation operator $\Pi_{p,q,r}$ as well as the computation of reconstruction kernels. Chapter 8 is entirely dedicated to the convergence and stability analysis of the method. Since the null space of $\mathbf{D}$ contains the potential fields ∇p with vanishing boundary values we introduce defect correction methods in Chapter 9. One feature of these methods is that we only need the measured data to compute the correction term and not the approximate solution.

6

A semi-discrete setup for Doppler tomography

In [50] and [120] the authors describe how ultrasound signals and the *Doppler effect* can be used to get informations along lines about a velocity field $\mathbf{f}$ of a moving fluid within a region $\Omega \subset \mathbb{R}^3$. This region Ω is supposed to be a bounded domain and represents the object under consideration. We give a brief summary of the derivation from the measurement setup to the mathematical model.

If we emit a signal $\tau(t) = e^{\imath k_0 t}$ with frequency k_0 along a line L which hits a particle of the fluid with velocity ν, then the frequency of the signal will be increased or decreased by the *Doppler shift*

$$\Delta = \frac{2\,c\,k_0\,\nu}{c^2 - \nu^2}\,,$$

where c denotes the velocity of sound within the medium. Note that it is necessary that the fluid contains particles causing a Doppler shift for this measurement procedure. Since usually $\nu \ll c$ the Doppler shift may be approximated by $\Delta \approx \kappa\,\nu$ with $\kappa = 2k_0/c$, which means that Δ is approximately proportional to the velocity ν of the particle causing the Doppler shift. Our experimental setup must satisfy that essential assumption. The signal $\sigma(t)$ which is received at the detector consists then of a superposition of reflections from particles along L contributing to the signal,

$$\sigma(t) = \frac{1}{2\pi}\int_{\mathbb{R}} e^{\imath\,(k_0+\kappa\,\nu)\,t}\,\mathrm{d}S(\mathbf{f},L,\nu)\,, \qquad \Delta = \kappa\,\nu\,. \tag{6.1}$$

Here, $\mathrm{d}S$ is a positive Radon measure characterized by

$$\mathrm{d}S(\mathbf{f},L,\nu) = \mathrm{meas}\{x \in L : \nu \le \langle \theta_L, \mathbf{f}(x)\rangle < \nu + \mathrm{d}\nu\}\,, \tag{6.2}$$

where $\theta_L \in S^2 = \{x \in \mathbb{R}^3 : \|x\| = 1\}$ is the vector of direction of L. The measure $\mathrm{d}S$ is called the *velocity spectrum* of $\mathbf{f}$ and can be interpretated as the number of particles with velocity ν along L. Defining

$$S(\mathbf{f}, L, \nu) = \text{meas}\{x \in L : \langle \theta_L, \mathbf{f}(x) \rangle < \nu\}$$

the velocity spectrum $\mathrm{d}S$ has the representation $\mathrm{d}S = S'\mathrm{d}\nu$, where S' is the derivative of S with respect to ν. Hence (6.1) essentially is the inverse Fourier transform of S' and we obtain the first moment of the velocity spectrum using Fourier techniques

$$\int_{\mathbb{R}} \nu \, dS(\mathbf{f}, L, \nu) = \int_{L \cap \Omega} \langle \theta_L, \mathbf{f}(x) \rangle \, \mathrm{d}\ell(x) =: \mathbf{D}f(L) \,. \tag{6.3}$$

We call the mapping $\mathbf{f} \mapsto \mathbf{D}\mathbf{f}(L)$ the *Doppler transform.*

In principle, L varies over all lines in $\mathbb{R}^3$. However, we confine to lines being parallel to one of the coordinate planes $\{x_j = 0\}$, $j = 1, 2, 3$, which corresponds to the measurement geometry suggested by JUHLIN [50]. This geometry scans the object slice by slice where the parallel geometry known from the 2D computerized tomography is applied in each slice. Once the slices parallel to one coordinate plane are scanned, the measurements device has to be turned by 90° and the procedure is repeated, i.e. the slices parallel to another coordinate plane are measured. This leads to three sets of data corresponding to the three planes $\{x_j = 0\}$, $j = 1, 2, 3$. The parallel geometry in two dimensions is illustrated in Figure 6.1.

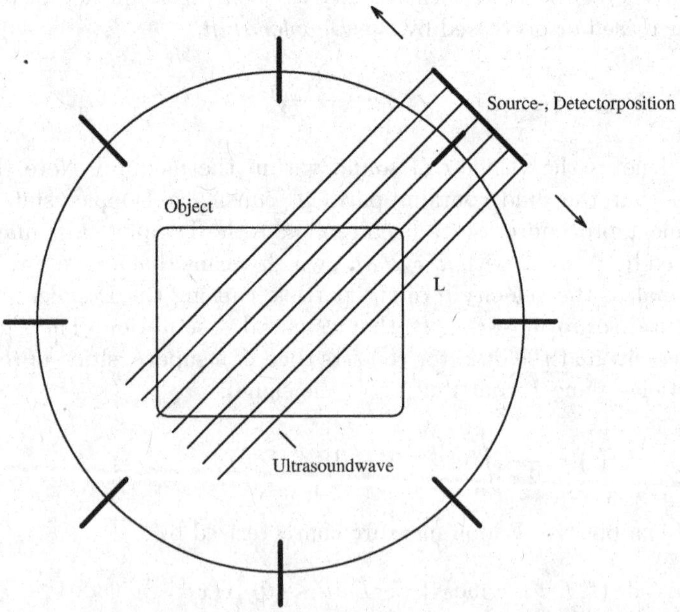

Fig. 6.1. The parallel geometry in two dimensions. Eight transducer/detector positions are displayed. The measurement device is shifted along the tangent of the scanning circle and ultrasound signals are emitted along parallel lines.

The first step to establish the Doppler transform as continuous operator between suitable L^2-spaces is to parameterize the lines L being parallel to $\{x_j = 0\}$. Before doing so, we introduce notations which seem to be awkward but which are necessary. Vectors will be written either horizontally or vertically depending on covenience. Let $w_1 = (0,0,1)$, $w_2 = (1,0,0)$ and $w_3 = (0,1,0)$ be a permutation of the standard unit vectors. To each w_j we associate embeddings $\mathcal{P}_j : \mathbb{R}^2 \to w_j^\perp$, $j = 1,2,3$, by $\mathcal{P}_1(x_1,x_2) = (x_1,x_2,0)$, $\mathcal{P}_2(x_1,x_2) = (0,x_1,x_2)$ and $\mathcal{P}_3(x_1,x_2) = (x_1,0,x_2)$. For the parameterization of the lines L we need three quantities: an angle $\varphi \in [0,2\pi]$ to define the direction of the line, the distance from the w_j-coordinate axis $s \in \mathbb{R}$ and the distance $a \in \mathbb{R}$ of the line L from the coordinate plane $w_j^\perp = \{x \in \mathbb{R}^3 : \langle x, w_j\rangle = 0\}$. With the help of these quantities, lines being parallel to $w_j^\perp$ can be defined by

$$L_j(\varphi,s,a) = \{x \in \mathbb{R}^3 : \langle x, \mathcal{P}_j\omega(\varphi)\rangle = s\,, \quad \langle x, w_j\rangle = a\}\,, \tag{6.4}$$

where $\omega(\varphi) = (\cos\varphi, \sin\varphi) \in S^1$ is the unit vector in $\mathbb{R}^2$ with polar angle φ. For $j = 1$ the meaning of the parameters is emphasized in Figure 6.2.

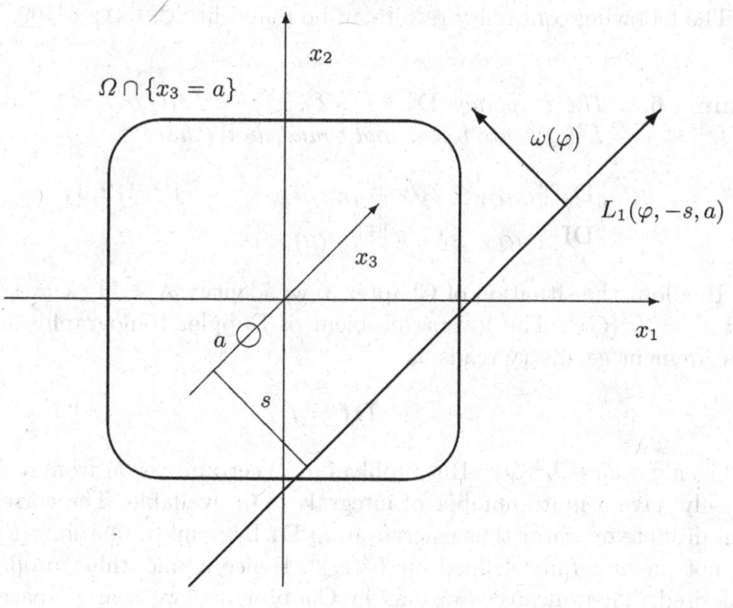

Fig. 6.2. Parameters of the first component of the Doppler transform $\mathbf{D}_1\mathbf{f}$.

From now on, we assume $\Omega = \Omega^3 = \{x \in \mathbb{R}^3 : \|x\| < 1\}$ to be the open unit ball in $\mathbb{R}^3$ if not indicated otherwise. If $\mathbf{f}$ is compactly supported, then this can always be accomplished by a suitable re-scaling of $\mathbf{f}$. In (6.3) the integration is now to be taken along lines L_j of the form (6.4). The vectors of direction θ_{L_j} of such lines L_j are given by $w_j \times \mathcal{P}_j\omega(\varphi)$. Applying this to (6.3), then the 3D-*Doppler transform* is defined in correspondence to our special measure geometry as a mapping

$$\mathbf{D} = (\mathbf{D}_1, \mathbf{D}_2, \mathbf{D}_3) : L^2(\Omega^3, \mathbb{R}^3) \to L^2(Q)^3, \qquad Q = [0, 2\pi] \times [-1, 1]^2, \tag{6.5}$$

where

$$\mathbf{D}_j \mathbf{f}(\varphi, s, a) = \int_{L_j(\varphi,s,a)\cap\Omega^3} \langle w_j \times \mathcal{P}_j\omega(\varphi), \mathbf{f}(x)\rangle \, \mathrm{d}\ell(x), \qquad j = 1, 2, 3. \tag{6.6}$$

Here, $L^2(\Omega^3, \mathbb{R}^3) = \{\mathbf{f} : \Omega^3 \to \mathbb{R}^3 : \int_{\Omega^3} \|\mathbf{f}(x)\|^2 \, \mathrm{d}x < \infty\}$ is the Hilbert space of all square integrable vector fields on Ω^3 and $L^2(Q)^3 = L^2(Q) \times L^2(Q) \times L^2(Q)$ is the three-fold cartesian product of the space $L^2(Q)$ with inner product

$$\langle g, h\rangle_{L^2(Q)^3} = \sum_{i=1}^{3} \int_{-1}^{1} \int_{0}^{2\pi} \int_{-1}^{1} g_i(\varphi, s, a)\, h_i(\varphi, s, a) \, \mathrm{d}s \, \mathrm{d}\varphi \, \mathrm{d}a, \quad g, h \in L^2(Q)^3.$$

The following continuity result can be found in SCHUSTER [109, Theorem 2.3].

Lemma 6.1. *The mappings* $\mathbf{D}_j : L^2(\Omega^3, \mathbb{R}^3) \to L^2(Q)$, $j = 1, 2, 3$ *and* $\mathbf{D} : L^2(\Omega^3, \mathbb{R}^3) \to L^2(Q)^3$ *are linear and bounded. We have*

$$\|\mathbf{D}_j \mathbf{f}\|_{L^2(Q)} \le 2\sqrt{\pi} \, \|\mathbf{f}\|_{L^2(\Omega^3,\mathbb{R}^3)}, \qquad j = 1, 2, 3,$$
$$\|\mathbf{D}\mathbf{f}\|_{L^2(Q)^3} \le 6\sqrt{\pi} \, \|\mathbf{f}\|_{L^2(\Omega^3,\mathbb{R}^3)}.$$

To adopt the situation of Chapter 3, we identify $\mathbf{A} = \mathbf{D}$, $X = L^2(\Omega^3, \mathbb{R}^3)$ und $Y = L^2(Q)^3$. The inverse problem of Doppler tomography in Juhlin's measurement geometry reads as

$$\mathbf{D} f = g \tag{6.7}$$

for given data $g \in L^2(Q)^3$. But problem (6.7) certainly is far from reality, since we only have a finite number of integrals (6.6) available. The corresponding semi-discrete operator thus emerges from $\mathbf{Df}$ by point evaluations in Q. These are not meaningfully defined on $L^2(Q)^3$. Hence a smoothing property of $\mathbf{D}$ is desired. That means – just as in Chapter 3 – we search spaces $X_1 \hookrightarrow L^2(\Omega^3, \mathbb{R}^3)$ and $Y_1 \hookrightarrow L^2(Q)^3$ with continuous and dense embeddings such that $\mathbf{D} : X_1 \to Y_1$ is bounded. Such a smoothing property exists and has been proven in [109, Theorem 2.10]. Before stating the result (Theorem 6.2) we continue by defining appropriate function spaces.

For real numbers $\alpha, \beta \ge 0$ and $j = 1, 2, 3$ we introduce anisotropic Sobolev spaces $\mathcal{X}_j^{\alpha,\beta}$ as the closure of the Schwartz space $\mathcal{D}(\Omega^3) = C_0^\infty(\Omega^3)$ with respect to the norms

$$\|v\|_{\mathcal{X}_j^{\alpha,\beta}} = \left(\int_{\mathbb{R}^3} (1+\xi_1^2+\xi_2^2)^\alpha (1+\xi_3^2)^\beta \left| \hat{v}\Big(\mathcal{P}_j(\xi_1,\xi_2)+\xi_3\, w_j\Big) \right|^2 \mathrm{d}\xi \right)^{1/2} .$$

The expression $\mathcal{P}_j(\xi_1,\xi_2)+\xi_3\, w_j$ is a permutation of the entries of $\xi = (\xi_1,\xi_2,\xi_3)$. E.g. we have

$$\|v\|_{\mathcal{X}_1^{\alpha,\beta}} = \left(\int_{\mathbb{R}^3} (1+\xi_1^2+\xi_2^2)^\alpha (1+\xi_3^2)^\beta \, |\hat{v}(\xi)|^2 \, \mathrm{d}\xi \right)^{1/2} .$$

The Sobolev spaces $\mathcal{X}_j^{\alpha,\beta}$ are called anisotropic, since the parameters α and β allow for different orders of smoothing in the corresponding variables. For $\alpha \geq 0$ and an open domain $G \subset \mathbb{R}^m$, let $H^\alpha(G)$ be the Sobolev space of order α and $H_0^\alpha(G)$ be the H^α-closure of the set of all tempered distributions that are compactly supported in G. A detailed outline of the theory of Sobolev spaces can be found e.g. in the books of ADAMS [3] and MAZ'JA [77].

Theorem 6.2. *The Doppler transform* $\mathbf{D}$ *maps the Cartesian product* $(\mathcal{X}_1^{\alpha,\beta} \cap \mathcal{X}_3^{\alpha,\beta}) \times (\mathcal{X}_1^{\alpha,\beta} \cap \mathcal{X}_2^{\alpha,\beta}) \times (\mathcal{X}_2^{\alpha,\beta} \cap \mathcal{X}_3^{\alpha,\beta})$ *continuously to the tensor product space* $\big(H^{\alpha+1/2}(Z) \hat{\otimes} H_0^\beta(-1,1)\big)^3$ *with* $Z = (0,2\pi) \times (-1,1)$. *There exist constants* $c_j > 0$, $j = 1,2,3$, *validating the estimates*

$$\|\mathbf{D}_j f\|_{H^{\alpha+1/2}(Z)\hat{\otimes} H_0^\beta(-1,1)} \leq c_j \left(\|f_j\|_{\mathcal{X}_j^{\alpha,\beta}} + \|f_{j+1}\|_{\mathcal{X}_j^{\alpha,\beta}} \right), \quad j = 1,2,$$
$$\|\mathbf{D}_3 f\|_{H^{\alpha+1/2}(Z)\hat{\otimes} H_0^\beta(-1,1)} \leq c_3 \left(\|f_1\|_{\mathcal{X}_j^{\alpha,\beta}} + \|f_3\|_{\mathcal{X}_j^{\alpha,\beta}} \right).$$

Theorem 6.2 says that $\mathbf{D}$ smoothes in two variables by the factor 1/2, whereas it has no smoothing property with respect to the remaining third variable. This makes perfect sense in view of Definition (6.6), since $\mathbf{D}_j f$ acts on one variable as the identity mapping, which is a consequence of our particular measure geometry.

If we write

$$\mathcal{X}^{\alpha,\beta} := (\mathcal{X}_1^{\alpha,\beta} \cap \mathcal{X}_3^{\alpha,\beta}) \times (\mathcal{X}_1^{\alpha,\beta} \cap \mathcal{X}_2^{\alpha,\beta}) \times (\mathcal{X}_2^{\alpha,\beta} \cap \mathcal{X}_3^{\alpha,\beta})$$

and

$$\mathcal{Y}^{\alpha,\beta} := H^{\alpha+1/2}(Z) \hat{\otimes} H_0^\beta(-1,1),$$

then Theorem 6.2 simply says that

$$\mathbf{D}_j : \mathcal{X}^{\alpha,\beta} \to \mathcal{Y}^{\alpha,\beta} \tag{6.8}$$

is linear and bounded. In analogy with Chapter 3, we set $X_1 = \mathcal{X}^{\alpha,\beta}$ and $Y_1 = \mathcal{Y}^{\alpha,\beta}$. We note that $\mathcal{X}^{0,0} = L^2(\Omega^3, \mathbb{R}^3)$.

It is now time to define the observation operator, which describes the measurement process. In practical situations only a finite number of measurements are available. Hence, we set the observation operator to be point evaluations of the data sets $\mathbf{D}_j\mathbf{f}$. One can show that point evaluations are continuous functionals on $H^t(\mathbb{R}^m)$ only for $t > m/2$. Hence Theorem 6.2 assures that point evaluations of $\mathbf{D}_j\mathbf{f}$ are well-defined if only α and β are greater than 1/2. At first we have to specify a sampling scheme (φ_l, s_i, a_k). We choose the equidistant sampling

$$\begin{aligned} \varphi_l &= l \cdot h_\varphi\,, & h_\varphi &= 2\pi/p\,, & l &= 0,\dots,p-1\,,\\ s_i &= i \cdot h_s\,, & h_s &= 1/q\,, & i &= -q,\dots,q-1\,,\\ a_k &= k \cdot h_a\,, & h_a &= 1/r\,, & k &= -r,\dots,r-1 \end{aligned}$$

with integers $p, q, r \in \mathbb{N}$. The observation operator evaluates $\mathbf{D}_j\mathbf{f}$ at (φ_l, s_i, a_k). For $\alpha, \beta > 1/2$ point evaluations are continuous on $H^{\alpha+1/2}(Z)$ and $H_0^\beta(-1,1)$, respectively. Thus, the mappings

$$\Psi_{p,q} : H^{\alpha+1/2}(Z) \to \mathbb{R}^{2pq}\,, \qquad (\Psi_{p,q}v)_{l,i} = v(\varphi_l, s_i)\,, \tag{6.9}$$

and

$$\Psi_r : H_0^\beta(-1,1) \to \mathbb{R}^{2r}\,, \qquad (\Psi_r v)_k = v(a_k)$$

represent bounded functionals. The tensor product

$$\Psi_{p,q,r} := \Psi_{p,q} \otimes \Psi_r : \mathcal{Y}^{\alpha,\beta} \to \mathbb{R}^{2pq} \otimes \mathbb{R}^{2r} = \mathbb{R}^n\,, \quad n = 4pqr$$

acts continuously on $\mathcal{Y}^{\alpha,\beta}$ and serves as observation operator for the introduced model of Doppler tomography. The inverse problem now reads as: To given measurements $g_{p,q,r} \in \mathbb{R}^{3n}$, which are possibly perturbed by noise, find a vector field $\mathbf{f} \in \mathcal{X}^{\alpha,\beta}$, $\alpha, \beta > 1/2$, satisfying

$$\Psi_{p,q,r}\,\mathbf{Df} = g_{p,q,r}\,, \tag{6.10}$$

where $\Psi_{p,q,r}$ acts on $\mathbf{Df}$ as

$$\Psi_{p,q,r}\,\mathbf{Df} = (\Psi_{p,q,r}\,\mathbf{D}_1\mathbf{f}, \Psi_{p,q,r}\,\mathbf{D}_2\mathbf{f}, \Psi_{p,q,r}\,\mathbf{D}_3\mathbf{f})\,.$$

Following the lines in the proof of RIEDER, SCHUSTER [102, Theorem 5.1] we deduce that $\Psi_{p,q,r}\,\mathbf{D}$ can not be extended continuously on $L^2(\Omega^3, \mathbb{R}^3)$.

Lemma 6.3. *The semi-discrete Doppler transform*

$$\Psi_{p,q,r}\,\mathbf{D} : \mathcal{X}^{\alpha,\beta} \subset L^2(\Omega^3, \mathbb{R}^3) \to \mathbb{R}^{3n}$$

is unbounded with respect to the L^2-norm topology for all real numbers $\alpha, \beta > 1/2$. In other words: The mapping $\Psi_{p,q,r}\,\mathbf{D}$ has no bounded extension to $L^2(\Omega^3, \mathbb{R}^3)$.

The adjoint operator belonging to $\Psi_{p,q,r}\,\mathbf{D}$ does not exist due to Lemma 6.3. Moreover, one can show that $\mathsf{D}((\Psi_{p,q,r}\,\mathbf{D})^*) = \{0\}$. In order to use the method of approximate inverse for solving (6.10) according to the concepts presented in Chapter 3, we have to fix the reconstruction kernels by means of the observation operator $\Psi_{p,q,r}$ and an analytic kernel for $\mathbf{D}$ in analogy to (3.21). This process is being specified in Chapter 7.

Remark 6.4. In contrast to the Radon transform, the Doppler transform (6.3) is not injective. When we integrate over all lines in $\mathbb{R}^3$, the null space of $\mathbf{D}$ consists of all potential fields with vanishing boundary values, that means all fields $\mathbf{f} = \nabla v$ with $v \in H_0^1(\Omega^3)$. Thus, only the solenoidal part of $\mathbf{f}$ can be recovered from $\mathbf{Df}$. A proof of this well-known fact can be found e.g. in the book of SHARAFUTDINOV [116]. Inversion formulas yielding the solenoidal part of $\mathbf{f}$ have been derived e.g. by DENISJUK [21], SHARAFUTDINOV [116], see also NATTERER, WÜBBELING [84, Theorem 2.27]. However, the attenuated vectorial Radon transform as a matter of fact *is* injective, see Bukgheim, Kazantsev [13] and NATTERER [82]. The null space of $\mathbf{D}_j$ is specified in [110]. E.g. we have

$$\mathsf{N}(\mathbf{D}_1) = \left\{(\partial_{x_1} v, \partial_{x_2} v, w) : v \in H_0^1(\Omega^3), \quad w \in L^2(\Omega^3)\right\}.$$

7

Solving the semi-discrete problem

In this chapter we formulate the semi-discrete approximate inverse (3.17) to solve the inverse problem (6.10). We first have to state concrete representations for the interpolation operator $\Pi_{p,q,r}$ associated with $\Psi_{p,q,r}$ and a mollifier operator $\mathbf{E}_d$ according to the abstract formulations (3.22) and (3.7) which is done in Section 7.1. Section 7.1 also contains the proof of certain invariance properties of the adjoint $\mathbf{D}_j^*$, which we use to improve the efficiency of our algorithm. Section 7.2 includes a scheme for the computation of reconstruction kernels for the mappings $\mathbf{D}_j$. The last section finally summarizes the inversion method and shows some reconstructions from synthetic data.

7.1 Definition of the operators $\Pi_{p,q,r}$ and $\mathbf{E}_d$

We start by constructing an interpolation operator $\Pi_{p,q,r}$, which is associated with the observation operator $\Psi_{p,q,r}$. As Riesz system $\{\phi_k\} \subset L^2(Q)$ we take tensor products of piecewise constant B-splines. More explicitly, let S_φ, S_s and S_a be the spaces of piecewise constant B-splines corresponding to the sets of nodes $\{\varphi_l\}$, $\{s_i\}$ and $\{a_k\}$, respectively. Bases of these spaces are given by

$$\begin{aligned} b_{p,l}^{(0)} &= \chi_{[\varphi_l,\varphi_{l+1})}\,, & 0 \le l \le p-1\,, \\ b_{q,i}^{(0)} &= \chi_{[s_i,s_{i+1})}\,, & -q \le i \le q-1\,, \\ b_{r,k}^{(0)} &= \chi_{[a_k,a_{k+1})}\,, & -r \le k \le r-1\,, \end{aligned}$$

where χ_I always means the characteristic function of an interval I. Hence, the tensor products

$$\{b_{p,j}^{(0)} \otimes b_{q,i}^{(0)} \otimes b_{r,k}^{(0)} : 0 \le l \le p-1\,,\ -q \le i \le q-1\,,\ -r \le k \le r-1\} \quad (7.1)$$

form a basis of $V_{p,q,r} = S_\varphi \otimes S_s \otimes S_a$. The Riesz property (3.6) of (7.1) is proved by simple calculations. In analogy with our abstract concept (3.22), we define the interpolation operator $\Pi_{p,q,r} : \mathcal{Y}^{\alpha,\beta} \to V_{p,q,r} \subset L^2(Q)$ for real $\alpha, \beta > 1/2$ by

$$\Pi_{p,q,r} v := \sum_{l=0}^{p-1} \sum_{i=-q}^{q-1} \sum_{k=-r}^{r-1} (\Psi_{p,q,r} v)_{l,i,k} \, b_{p,l}^{(0)} \otimes b_{q,i}^{(0)} \otimes b_{r,k}^{(0)} , \qquad v \in \mathcal{Y}^{\alpha,\beta} . \quad (7.2)$$

We have to verify the important approximation property (3.23) and boundedness property (3.24).

Lemma 7.1. *Let $\alpha, \beta > 1/2$. The mapping $\Pi_{p,q,r} : \mathcal{Y}^{\alpha,\beta} \to V_{p,q,r}$ satisfies the approximation property*

$$\|\Pi_{p,q,r} v - v\|_{L^2(Q)} \le C_\Pi \, \rho \, \|v\|_{\mathcal{Y}^{\alpha,\beta}} , \qquad v \in \mathcal{Y}^{\alpha,\beta} \quad (7.3)$$

for a constant $C_\Pi > 0$ and

$$\rho = \rho(\beta, h_\varphi, h_s, h_a) = \max\{h_\varphi, h_s\} + h_a^{\min\{\beta,1\}} . \quad (7.4)$$

Furthermore, we have the uniform boundedness

$$\|\Pi_{p,q,r} v\|_{L^2(Q)} \le C_\mathrm{b} \, \|v\|_{\mathcal{Y}^{\alpha,\beta}} , \qquad v \in \mathcal{Y}^{\alpha,\beta} \quad (7.5)$$

where $C_\mathrm{b} > 0$.

Proof. To a real number $\kappa > 0$ we define $\Pi_{p,q} : H^{\kappa+1}(Z) \to S_\varphi \otimes S_s$ and $\Pi_r : H^{\kappa+1/2}(-1,1) \to S_a$ by

$$\Pi_{p,q} v := \sum_{l=0}^{p-1} \sum_{i=-q}^{q-1} (\Psi_{p,q} v)_{l,i} \, b_{p,l}^{(0)} \otimes b_{q,i}^{(0)} \quad \text{und} \quad \Pi_r v := \sum_{k=-r}^{r-1} (\Psi_r v)_k \, b_{r,k}^{(0)} .$$

The interpolation operator $\Pi_{p,q,r}$ can then be expressed as the tensor product $\Pi_{p,q,r} = \Pi_{p,q} \otimes \Pi_r$. Using results from approximation theory for B-splines, see e.g. SCHUMAKER [108, Chapter 12], we derive constants $c_1, c_2, C_1, C_2 > 0$ which do not depend on p, q and r such that the estimates

$$\|\Pi_{p,q}\|_{H^{\kappa+1}(Z) \to L^2(Z)} \le c_1 \quad \text{und} \quad \|\Pi_r\|_{H^{\kappa+1/2}(Z) \to L^2(-1,1)} \le c_2 , \quad (7.6)$$

as well as

$$\|I - \Pi_{p,q}\|_{H^{\kappa+1}(Z) \to L^2(Z)} \le C_1 \max\{h_\varphi, h_s\} \quad (7.7)$$

and

$$\|I - \Pi_r\|_{H^{\kappa+1/2}(-1,1) \to L^2(-1,1)} \le C_2 \, h_a^{\min\{\kappa+1/2,1\}} \quad (7.8)$$

are valid. Since $\Pi_{p,q,r}$ is the tensor product of $\Pi_{p,q}$ and Π_r, its norm is estimated as

$$\|\Pi_{p,q,r}\|_{\mathcal{Y}^{\alpha,\beta} \to L^2(Q)} \le \|\Pi_{p,q}\|_{H^{\alpha+1/2}(Z) \to L^2(Z)} \, \|\Pi_r\|_{H^\beta(-1,1) \to L^2(-1,1)} \le c_1 \, c_2 ,$$

according to AUBIN [6, Prop. 12.4.1]. This is (7.5) with $C_b = c_1 c_2$. The approximation property (7.3) is obtained from

$$\begin{aligned}
&\|I - \Pi_{p,q,r}\|_{\mathcal{Y}^{\alpha,\beta}\to L^2(Q)} = \|I \otimes I - \Pi_{p,q,r}\|_{\mathcal{Y}^{\alpha,\beta}\to L^2(Q)}\\
&\le \|I \otimes (I - \Pi_{p,q})\|_{\mathcal{Y}^{\alpha,\beta}\to L^2(Q)} + \|\Pi_{p,q} \otimes (I - \Pi_r)\|_{\mathcal{Y}^{\alpha,\beta}\to L^2(Q)}\\
&\le C_{\mathcal{Y}^{\alpha,\beta},L^2(Q)}\, \|I - \Pi_{p,q}\|_{H^{\alpha+1/2}(Z)\to L^2(Z)} + c_1\, \|I - \Pi_r\|_{H^\beta(-1,1)\to L^2(-1,1)}\\
&\le C_{\mathcal{Y}^{\alpha,\beta},L^2(Q)}\, C_1 \max\{h_\varphi, h_s\} + c_1\, C_2\, h_a^{\min\{\kappa+1/2,1\}}
\end{aligned}$$

where we applied the estimates (7.6), (7.7) and (7.8). The constant $C_{\mathcal{Y}^{\alpha,\beta},L^2(Q)}$ is equal to the norm of the embedding $\mathcal{Y}^{\alpha,\beta} \hookrightarrow L^2(Q)$. Finally we set $C_\Pi = \max\{c_1 C_2, C_{\mathcal{Y}^{\alpha,\beta},L^2(Q)} C_1\}$. □

Before we state the mollifier operator $\mathbf{E}_d$, we prove the existence of an intertwining relation for $\mathbf{D}_j^*$ which fulfills all conditions of Lemma 3.9. To this end, we define for $d > 0$ and $k \in \mathbb{Z}^3$ the mappings

$$\mathcal{T}_j^{d,k} f := d^3 f(d\,x - k)$$

which act on $L^2(\mathbb{R}^3)$ and

$$\mathcal{G}_j^{d,k} g(\varphi, s, a) := d^3\, g\big(\varphi, d\,s - \langle \mathcal{P}_j^* k, \omega(\varphi)\rangle, d\,a - \langle k, w_j\rangle\big)$$

acting on $L^2([0, 2\pi] \times \mathbb{R}^2)$.

Lemma 7.2. *Let* $\mathbf{D}_j^* : L^2([0, 2\pi] \times \mathbb{R}^2) \to L^2(\mathbb{R}^3)^3$ *be the adjoint operator of* $\mathbf{D}_j$ *with respect to the given* L^2*-spaces. Then*

$$\mathbf{D}_j^*\, \mathcal{G}_j^{d,k} = \mathcal{T}_j^{d,k}\, \mathbf{D}_j^*\,, \qquad d > 0\,,\ k \in \mathbb{Z}^3\,. \tag{7.9}$$

Proof. The adjoint operator $\mathbf{D}_j^* : L^2([0, 2\pi] \times \mathbb{R}^2) \to L^2(\mathbb{R}^3)^3$ has the representation

$$\begin{aligned}
\mathbf{D}_j^* g(x) = \mathcal{P}_j\Bigg(&\iota_j \int_0^{2\pi} g\big(\varphi, \langle \mathcal{P}_j^* x, \omega(\varphi)\rangle, \langle x, w_j\rangle\big) \sin\varphi\, d\varphi\,,\\
&-\iota_j \int_0^{2\pi} g\big(\varphi, \langle \mathcal{P}_j^* x, \omega(\varphi)\rangle, \langle x, w_j\rangle\big) \cos\varphi\, d\varphi\Bigg)\,,
\end{aligned}$$

where $\iota_1 = \iota_2 = -1$ und $\iota_3 = 1$. This result can be found in [109, Formula (2.10)]. The invariance property (7.9) is now obtained in the same way as in the proof of the corresponding Lemma 2.7 for the Radon transform by a straightforward computation. □

The intertwining (7.9) suggests to generate a mollifier for $\mathbf{D}_j$ by means of the mappings $\mathcal{T}_j^{d,k}$. Having functions $e^j \in L^2(\mathbb{R}^3)$ satisfying $\int e^j(x)\,dx = 1$ for $j = 1, 2, 3$ at hand, we therefore define mollifiers

$$e^j_{d,k}\,\delta_j := (\mathcal{T}_j^{d,k} e^j)\,\delta_j\,, \tag{7.10}$$

where $\delta_j \in \mathbb{R}^3$ are the standard unit vectors[1]. Note that $d > 0$ represents the regularization parameter. This fact implies for large values $d > 0$ the approximation

$$\langle \mathbf{f}, e^j_{d,k}\,\delta_j\rangle_{L^2(\mathbb{R}^3)^3} = \langle \mathbf{f}_j, e^j_{d,k}\rangle_{L^2(\mathbb{R}^3)} \approx \mathbf{f}_j(d^{-1}\,k)\,, \quad k \in \mathbb{Z}^3\,.$$

Thus, the reconstruction points are given by $d^{-1}\,k$. These are contained in Ω^3 when $\|k\| < d$.

From Lemma 3.9 we deduce that the reconstruction kernels $v^j_{d,k}$ associated with $e^j_{d,k}$ can be generated by an application of $\mathcal{G}_j^{d,k}$.

Corollary 7.3. *Assume that* $d \geq 1$ *and* $k \in \mathbb{Z}^3$ *with* $\|k\| \leq d - 1$. *If* $v^j \in L^2([0, 2\,\pi] \times \mathbb{R}^2)$ *satisfies*

$$\mathbf{D}_j^* v^j = \mathcal{P}_{\mathsf{N}(\mathbf{D}_j)^\perp} e^j\,\delta_j\,,$$

then

$$v^j_{d,k} = \mathcal{G}_j^{d,k} v^j \tag{7.11}$$

is a reconstruction kernel belonging to $e^j_{d,k}$. *That means,* $v^j_{d,k}$ *solves*

$$\mathbf{D}_j^* v^j_{d,k} = \mathcal{P}_{\mathsf{N}(\mathbf{D}_j)^\perp}(e^j_{d,k}\,\delta_j)\,. \tag{7.12}$$

Proof. A simple calculation shows

$$\|\mathcal{T}_j^{d,k} f\|_{L^2(\mathbb{R}^3)} = d^{3/2}\,\|f\|_{L^2(\mathbb{R}^3)}$$

and

$$(\mathcal{G}_j^{d,k})^{-1} g(\varphi, s, a) = d^{-3}\,g\Big(\varphi, d^{-1}\,(s + \langle \mathcal{P}_j^* k, \omega(\varphi)\rangle), d^{-1}\,(a + \langle k, w_j\rangle)\Big)$$

proving that $\mathcal{T}_j^{d,k}$ is the multiple of an isometry and $\mathcal{G}_j^{d,k}$ is invertible and hence onto. Putting $\mathbf{A} = \mathbf{D}_j$, $X = L^2_{\Omega^3}(\mathbb{R}^3) := \{f \in L^2(\mathbb{R}^3) : \operatorname{supp} f \subset \overline{\Omega^3}\}$, $Y = L^2([0, 2\,\pi] \times \mathbb{R}^2)$, $S = \mathcal{G}_j^{d,k}$ and $T = \mathcal{T}_j^{d,k}$, then (7.11) readily follows from Lemma 3.9. The conditions for d and k assure that $\mathcal{T}_j^{d,k}(L^2_{\Omega^3}(\mathbb{R}^3)) \subset L^2_{\Omega^3}(\mathbb{R}^3)$. □

Remark 7.4. Note that Theorem 2.6 could not have been applied to prove (7.11). One reason is that $\mathbf{D}_j$ is not one-to-one, another reason is that we do not have an intertwining relation for $\mathbf{D}_j\,\mathbf{D}_j^*$ available.

[1] $\delta_1 = (1, 0, 0)$, $\delta_2 = (0, 1, 0)$, $\delta_3 = (0, 0, 1)$.

The last missing ingredients to define the mollifier operator $\mathbf{E}_d$ are specifications of a Riesz system $\{B_{d,k}\} \subset L^2(\Omega^3, \mathbb{R}^3)$ and of the mollifiers $e^j \in L^2(\mathbb{R}^3)$. Similar to the setting of $\Pi_{p,q,r}$, we define the functions $B_{d,k}$ with the help of tensor product-splines. Here, we use piecewise linear splines. Let $B = b_l \otimes b_l \otimes b_l$ be the tensor product of the first order B-spline $b_l = \chi_{[-1/2,1/2]} * \chi_{[-1/2,1/2]}$ [2], which we have already used in Section 3.1. Then, $B_{d,k}$ are defined as translated and dilated versions of B

$$B_{d,k}(x) = B(d\,x - k)\,, \qquad d > 0\,, \quad k \in \mathbb{Z}^3\,. \tag{7.13}$$

The family $\{B_{d,k}\}$ forms a Riesz system in $L^2(\Omega^3, \mathbb{R}^3)$. The mollifier operator $\mathbf{E}_d : L^2(\Omega^3, \mathbb{R}^3) \to L^2(\Omega^3, \mathbb{R}^3)$ finally reads as

$$\begin{aligned}(\mathbf{E}_d)_j \mathbf{f}(x) &:= \sum_{k \in \mathbb{Z}^3} \langle \mathbf{f}_j, e^j_{d,k} \rangle_{L^2(\Omega^3)}\, B_{d,k}(x) \\ &= \sum_{k \in \mathbb{Z}^3} \langle \mathbf{f}, e^j_{d,k}\, \delta_j \rangle_{L^2(\Omega^3, \mathbb{R}^3)}\, B_{d,k}(x)\,.\end{aligned} \tag{7.14}$$

In order to prove the mollifier property (3.8) of $\mathbf{E}_d$, we have to specify the particular choice of $e^j \in L^2(\mathbb{R}^3)$. When defining e^j, we want to pay tribute to the special structure of $\mathbf{D}_j$ as we did it in the definition of the space $\mathcal{Y}^{\alpha,\beta}$. For instance, $\mathbf{D}_1$ essentially acts as a two-dimensional Radon transform with respect to the variables (x_1, x_2) and as the identity with respect to x_3. In the same way, $\mathbf{D}_2$ leaves x_1 constant and $\mathbf{D}_3$ the variable x_2. This is to be taken into account when fixing e^j. Correspondingly, we define e^j as tensor products

$$\left.\begin{aligned} e^1(x) &:= \mathrm{p}(x_1, x_2)\, \mathrm{q}(x_3)\,, \\ e^2(x) &:= \mathrm{p}(x_2, x_3)\, \mathrm{q}(x_1)\,, \\ e^3(x) &:= \mathrm{p}(x_1, x_3)\, \mathrm{q}(x_2)\,. \end{aligned}\right\} \tag{7.15}$$

The functions p and q are chosen to be the mollifiers with compact support (2.5). For $\nu \in \mathbb{N}$ let

$$\mathrm{p}(s,t) = \mathrm{p}_\nu(s,t) := \frac{\nu+1}{\pi} \begin{cases} (1 - s^2 - t^2)^\nu\,, & s^2 + t^2 \le 1\,, \\ 0\,, & \text{else} \end{cases} \tag{7.16}$$

and

$$\mathrm{q}(s) = \mathrm{q}_\nu(s) := \frac{(2\,\nu+1)!!}{2^{\nu+1}\,\nu!} \begin{cases} (1 - s^2)^\nu\,, & |s| \le 1\,, \\ 0\,, & \text{else}\,. \end{cases} \text{[3]} \tag{7.17}$$

Setting e^j as in (7.15), (7.16) and (7.17) we have that

$$\int_{\mathbb{R}^3} e^j(x)\, \mathrm{d}x = 1\,,$$

[2] The symbol $f * g$ always denotes the convolution of two functions f and g.
[3] $(2\,\nu+1)!! = 1 \cdot 3 \cdot 5 \cdot \ldots \cdot (2\,\nu+1)$

i.e. the e^j have normalized mean values. Unfortunately, the support of e^j is not the closed unit ball $\overline{\Omega^3}$ but the cylinder $\overline{\Omega^2} \times [-1,1]$, which is slightly larger than $\overline{\Omega^3}$. By the scaling $e^j(\cdot) := 2^{3/2}\, e^j(\sqrt{2}\cdot)$ we in fact would achieve that e^j is supported in $\overline{\Omega^3}$. Since, in the following, we will scale the e^j anyway, see (7.10), and for the sake of a better readability, we omit the scaling right now and consider e^j being elements of $\mathcal{X}_j^{\lambda,\lambda}$ as long as $\lambda < \nu + 1/2$. The smoothness of e^j can be proved by calculating their Fourier transform and studying the decay behavior. At last, we state the mollifier property of $\mathbf{E}_d$.

Theorem 7.5. *Let $\{e^j_{d,k}\}$ be the sequence of mollifiers according to (7.10), (7.15), (7.16) and (7.17) and $\{B_{d,k}\}$ be the Riesz system (7.13). The operator $\mathbf{E}_d : L^2(\Omega^3,\mathbb{R}^3) \to L^2(\Omega^3,\mathbb{R}^3)$ defined by (7.14) satisfies*

$$\lim_{d\to\infty} \|\mathbf{E}_d\mathbf{f} - \mathbf{f}\|_{L^2(\Omega^3,\mathbb{R}^3)} = 0 \qquad \text{for all } \mathbf{f} \in L^2(\Omega^3,\mathbb{R}^3)\,. \tag{7.18}$$

If $\mathbf{f}_j \in \mathcal{X}_j^{\alpha,\beta}$ for $j = 1,2,3$ and $\alpha,\beta > 1/2$, then furthermore the estimate

$$\|(\mathbf{E}_d)_j\mathbf{f} - \mathbf{f}_j\|_{L^2(\Omega^3)} \le C_{\mathrm{m}}\left(d^{-\min\{2,\alpha\}} + d^{-\min\{2,\beta\}}\right)\|\mathbf{f}_j\|_{\mathcal{X}_j^{\alpha,\beta}} \tag{7.19}$$

holds for a suitable constant $C_{\mathrm{m}} > 0$.

Proof. Similar to $\Pi_{p,q,r}$, we will use the specific tensor product structure of $\mathbf{E}_d$ and apply the results (3.13) and (3.14) from Section 3.1. To this end, we first introduce two mappings $E_d^{(i)} : L^2(\mathbb{R}^i) \to L^2(\mathbb{R}^i)$ for $i = 1,2$ and $d > 0$ via

$$E_d^{(2)}v := \sum_{k\in\mathbb{Z}^2} \langle v, \mathrm{p}_{d,k}\rangle_{L^2(\mathbb{R}^2)}\, (b_l \otimes b_l)(d\,\cdot - k)$$

and

$$E_d^{(1)}u := \sum_{l\in\mathbb{Z}} \langle u, \mathrm{q}_{d,l}\rangle_{L^2(\mathbb{R})}\, b_l(d\,\cdot - l)\,,$$

where b_l is again the univariate B-spline of first order with support in $[-1,1]$ and

$$\mathrm{p}_{d,k}(s,t) = d^2\,\mathrm{p}(d\,s - k_1, d\,t - k_2)\,, \qquad \mathrm{q}_{d,l}(s) = d\,\mathrm{q}(d\,s - l)$$

are translations and dilations of p (7.16) and q (7.17). From (3.13) we deduce

$$\|E_d^{(i)} - I\|_{H^\alpha(\mathbb{R}^i)\to L^2(\mathbb{R}^i)} \le c_i\, d^{-\min\{2,\alpha\}}\,, \quad \alpha \ge 0\,, \quad i = 1,2 \tag{7.20}$$

with $c_i > 0$, $i = 1,2$. In the same way as the local boundedness (3.12), we prove that $E_d^{(2)}$ is uniformly bounded. That means, there exists a constant $\tilde{c} > 0$ which does not depend on d, such that

$$\|E_d^{(2)} f\|_{L^2(\mathbb{R}^2)} \le \tilde{c}\,\|f\|_{H^\alpha(\mathbb{R}^2)}\,. \tag{7.21}$$

We notice that $\mathbf{E}_d$ can be written as

$$(\mathbf{E}_d)_j f = (E_d^{(2)} \otimes E_d^{(1)})(f_j \circ \mathcal{Q}_j)\,, \quad j = 1,2,3\,, \tag{7.22}$$

with the permutations $\mathcal{Q}_j(x_1, x_2, x_3) := \mathcal{P}_j(x_1, x_2) + x_3\, w_j$. Following the arguments in the proof of Lemma 7.1 together with (7.20) and (7.21) yields

$$\begin{aligned}
&\|E_d^{(2)} \otimes E_d^{(1)} - I\|_{H^\alpha(\mathbb{R}^2)\hat{\otimes} H^\beta(\mathbb{R}) \to L^2(\mathbb{R}^3)} \\
&\quad \le \tilde{c}\, \|E_d^{(1)} - I\|_{H^\beta(\mathbb{R}) \to L^2(\mathbb{R}^3)} + C_{H^\beta, L^2}\, \|E_d^{(2)} - I\|_{H^\alpha(\mathbb{R}^2) \to L^2(\mathbb{R}^3)} \\
&\quad \le c_1\, \tilde{c}\, d^{-\min\{2,\beta\}} + C_{H^\beta, L^2}\, c_2\, d^{-\min\{2,\alpha\}} \\
&\quad \le C_{\mathrm{m}} \left(d^{-\min\{2,\beta\}} + d^{-\min\{2,\alpha\}} \right),
\end{aligned}$$

where $C_{H^\beta, L^2} > 0$ is the norm of the continuous embedding $H^\beta(\mathbb{R}) \hookrightarrow L^2(\mathbb{R})$ and $C_{\mathrm{m}} := \max\{\tilde{c}\, c_1, C_{H^\beta, L^2}\, c_2\}$. Let $\mathbf{f} \in \mathcal{D}(\Omega^3, \mathbb{R}^3)$. Then, for $j = 1, 2, 3$, we have

$$\begin{aligned}
\|(\mathbf{E}_d)_j f - f_j\|_{L^2(\Omega^3)} &\le \|(\mathbf{E}_d)_j f - f_j\|_{L^2(\mathbb{R}^3)} \\
&= \|(\mathbf{E}_d)_j f \circ \mathcal{Q}_j - f_j \circ \mathcal{Q}_j\|_{L^2(\mathbb{R}^3)} \\
&= \|(E_d^{(2)} \otimes E_d^{(1)})(f_j \circ \mathcal{Q}_j) - f_j \circ \mathcal{Q}_j\|_{L^2(\mathbb{R}^3)} \\
&\le C_{\mathrm{m}} \left(d^{-\min\{2,\alpha\}} + d^{-\min\{2,\beta\}} \right) \|f_j \circ \mathcal{Q}_j\|_{H^\alpha(\mathbb{R}^2)\hat{\otimes} H^\beta(\mathbb{R})}\,,
\end{aligned}$$

where we made use of representation (7.22). Taking into account that

$$\|f_j \circ \mathcal{Q}_j\|_{H^\alpha(\mathbb{R}^2)\hat{\otimes} H^\beta(\mathbb{R})} = \|f_j\|_{\mathcal{X}_j^{\alpha,\beta}}$$

and that the Schwartz space $\mathcal{D}(\Omega^3)$ is dense in $\mathcal{X}_j^{\alpha,\beta}$, we derive estimate (7.19) and at last the mollifier property (7.18) by means of a density argument. □

Provided that we have a solution of

$$\mathbf{D}_j^* v^j = \mathcal{P}_{\mathsf{N}(\mathbf{D}_j)^\perp} e^j\, \delta_j \tag{7.23}$$

at hand, we are able to formulate the semi-discrete approximate inverse to $\mathbf{D}_j$ and $\mathbf{D}$ according to (3.17). The solution of (7.23) is subject of Section 7.2.

Remark 7.6. We used the fact that $\mathbf{D}_j$ acts as a Radon transform with respect to two variables, whereas it leaves the remaining variable constant to define the mappings $\Psi_{p,q,r}$, $\Pi_{p,q,r}$ und $\mathbf{E}_d$. The whole convergence and stability analysis of the method as well as the computation of reconstruction kernels relies on this observation. It becomes clear, why we did not set $w_j = \delta_j$ in (6.6), because in that case we would have $\mathbf{D}_j^* g \perp w_j$ for all $g \in L^2(Q)$ and $j = 1, 2, 3$. The equations (7.23) then would only be solved by 0; a reconstruction kernel would not exist at all.

7.2 Computation of reconstruction kernels for $\mathbf{D}_j$

We present a recipe to solve the equations (7.23), which is suited for arbitrary mollifiers e^j of type (7.15), though we use the particular representation (7.16), (7.17). The equations (7.23) are equivalent to the normal equations

$$\mathbf{D}_j \mathbf{D}_j^* v^j = \mathbf{D}_j(e^j \delta_j), \qquad j = 1, 2, 3. \tag{7.24}$$

Using the relation $\partial_s \mathbf{D}_j = \mathbf{R}(w_j \cdot \nabla \times (e^j \delta_j))$, which is proved in STRÅHLÉN [124] or SCHUSTER [109, Lemma 2.4], one can show that the solutions v^j of (7.24) satisfy

$$(\mathbf{R}^* \otimes I)\, \partial_s v^j = w_j \cdot \nabla \times (e^j \delta_j), \qquad j = 1, 2, 3. \tag{7.25}$$

We again refer to [109] for a proof. Here, $\mathbf{R}^*$ is the adjoint operator of the two-dimensional Radon transform (2.17). Let again be Ω^2 the open unit disk in $\mathbb{R}^2$.

We perform the following calculations for $j = 1$ only, the computation of v^2, v^3 is done accordingly. Hence, we focus at the solution of

$$(\mathbf{R}^* \otimes I)\, \partial_s v^1 = -\partial_{x_2} e^1, \tag{7.26}$$

which is (7.25), if $j = 1$. Our starting point to solve (7.26) is the inversion formula (2.19) for the two-dimensional Radon transform $\mathbf{R}$. Applying (2.19), we deduce for the derivative $\partial_s v^1$ the representation

$$\partial_s v^1 = -(2\pi)^{-1} (\Lambda \mathbf{R} \otimes I)\, \partial_{x_2} e^1 = -(2\pi)^{-1} (\Lambda \mathbf{R}\, \partial_{x_2} \mathrm{p}) \otimes \mathrm{q}.$$

By means of the Fourier slice theorem (2.26) $\Lambda \mathbf{R}\, \partial_{x_2} \mathrm{p}$ can be calculated explicitly,

$$\begin{aligned} \Lambda \mathbf{R}\, \partial_{x_2} \mathrm{p}(\varphi, s) &= \int_{\mathbb{R}} |\sigma|\, \mathbf{F}\{\partial_{x_2} \mathrm{p}\}(\sigma\, \omega(\varphi))\, \mathrm{e}^{\imath s \sigma}\, \mathrm{d}\sigma \\ &= \imath \sin\varphi \int_{\mathbb{R}} |\sigma|\, \sigma\, \hat{\mathrm{p}}(\sigma\, \omega(0))\, \mathrm{e}^{\imath s \sigma}\, \mathrm{d}\sigma \\ &= -2 \sin\varphi \int_0^\infty \sigma^2\, \hat{\mathrm{p}}(\sigma, 0)\, \sin(s\sigma)\, \mathrm{d}\sigma. \end{aligned} \tag{7.27}$$

This yields

$$\partial_s v^1(\varphi, s, a) = \pi^{-1} \sin\varphi\, \mathrm{q}(a) \int_0^\infty \sigma^2\, \hat{\mathrm{p}}(\sigma, 0)\, \sin(s\sigma)\, \mathrm{d}\sigma.$$

Representation (2.32) implies

$$\hat{\mathrm{p}}(\sigma,0)=\frac{(\nu+1)!}{2\,\pi}\left(\frac{2}{\sigma}\right)^{\nu+1}\,\mathrm{J}_{\nu+1}(\sigma)\,. \tag{7.28}$$

We may conclude that

$$\begin{aligned}
&\partial_s v^1(\varphi,s,a)\\
&\quad=\frac{1}{\pi^2}\,(\nu+1)!\,2^\nu\,\sin\varphi\,\mathrm{q}(a)\int_0^\infty\sigma^{1-\nu}\,\mathrm{J}_{\nu+1}(\sigma)\,\sin(s\,\sigma)\,\mathrm{d}\sigma\\
&\quad=\frac{1}{\pi^2}\,\sin\varphi\,\mathrm{q}(a)\begin{cases}4\,\nu\,(\nu+1)\,s\,{}_2\mathrm{F}_1(2,1-\nu;\frac{3}{2};s^2)\,, & |s|<1\,,\\ -s^{-3}\,{}_2\mathrm{F}_1(2,\frac{3}{2};\nu+2;s^{-2})\,, & |s|\geq 1\,,\end{cases}
\end{aligned} \tag{7.29}$$

where we further applied formula (6.699.1) from Gradshteyn, Rizhik [32]. So far, we know the derivative with respect to s of the searched for reconstruction kernel v^1. According to the fundamental theorem of calculus, v^1 can be written as

$$v^1(\varphi,s,a)=\mathrm{I}(\varphi,s,a)+h(\varphi,a)\,, \tag{7.30}$$

where

$$\mathrm{I}(\varphi,s,a):=\int_{-1}^{s}\partial_s v^1(\varphi,t,a)\,\mathrm{d}t \tag{7.31}$$

is an antiderivative of $\partial_s v^1$ and h is a function which does not depend on s. We seek explicit expressions for I and h. To this end, we determine v^1 to be the *unique* solution of the normal equation (7.24) in $\overline{\mathsf{R}(\mathbf{D}_1)}$. This is an essential assumption. Theorem 3.9 from [109] proves that the function system

$$\mathcal{M}:=\{\sin(k\,\varphi)\,,\ \cos(k\,\varphi):k\in\mathbb{N}\cup\{0\}\,,\ k\neq 1\}$$

is orthogonal to $\overline{\mathsf{R}(\mathbf{D}_1)}$. This implies

$$\int_0^{2\pi}v^1(\varphi,s,a)\,\sin(k\,\varphi)\,\mathrm{d}\varphi=\int_0^{2\pi}v^1(\varphi,s,a)\,\cos(k\,\varphi)\,\mathrm{d}\varphi=0$$

if $k\neq 1$. From (7.29) we immediately read that

$$\int_0^{2\pi}\mathrm{I}(\varphi,s,a)\,\sin(k\,\varphi)\,\mathrm{d}\varphi=\int_0^{2\pi}\mathrm{I}(\varphi,s,a)\,\cos(k\,\varphi)\,\mathrm{d}\varphi=0$$

if $k\neq 1$. In view of (7.30), this yields

$$\int_0^{2\pi}h(\varphi,a)\,\sin(k\,\varphi)\,\mathrm{d}\varphi=\int_0^{2\pi}h(\varphi,a)\,\cos(k\,\varphi)\,\mathrm{d}\varphi=0$$

if $k \neq 1$. Setting $\varphi = 0$ and $s = 0$ in (7.24) (for $j = 1$) and taking into account the symmetry $\mathrm{I}(-\varphi, -s, a) = -\mathrm{I}(\varphi, s, a)$ lead to

$$\int_0^{2\pi} h(\varphi, a) \cos\varphi \, \mathrm{d}\varphi = 0 .$$

Assume for the moment the Fourier coefficient $\int_0^{2\pi} h(\vartheta, a) \sin\vartheta \, \mathrm{d}\vartheta$ to be known, then h could be expressed by

$$h(\varphi, a) = \frac{1}{\pi} \Big\{ \int_0^{2\pi} h(\vartheta, a) \sin\vartheta \, \mathrm{d}\vartheta \Big\} \sin\varphi ,$$

since $\mathcal{M} \cup \{\sin\varphi, \cos\varphi\}$ builds a complete orthogonal system in $L^2(0, 2\pi)$. To calculate this very coefficient we again consider the normal equation (7.24) and evaluate it for $j = 1$ at $\varphi = \pi/2$ and $s = 0$. The right-hand side of (7.24) then turns to

$$\mathbf{D}_1(e^1 \delta_1)(\pi/2, 0, a) = -\mathrm{q}(a) \int_{-1}^{1} (1 - t^2)^\nu \, \mathrm{d}t = -\frac{2^{\nu+1} \nu!}{(2\nu + 1)!!} \mathrm{q}(a)$$

whereas the left-hand side reads as

$$\begin{aligned} \mathbf{D}_1 \mathbf{D}_1^* v^1(\pi/2, 0, a) &= \mathbf{D}_1 \mathbf{D}_1^* \mathrm{I}(\pi/2, 0, a) + \mathbf{D}_1 \mathbf{D}_1^* h(\pi/2, 0, a) \\ &= \int_0^{2\pi} \int_{-1}^{1} \mathrm{I}(\vartheta, t \cos\vartheta, a) \sin\vartheta \, dt \, \mathrm{d}\vartheta + 2 \int_0^{2\pi} h(\vartheta, a) \sin\vartheta \, \mathrm{d}\vartheta . \end{aligned}$$

Hence, we get

$$\begin{aligned} h(\varphi, a) = &-\frac{1}{2\pi} \sin\varphi \\ &\times \left(\frac{2^{\nu+1} \nu!}{(2\nu + 1)!!} \mathrm{q}(a) + \int_0^{2\pi} \int_{-1}^{1} \mathrm{I}(\vartheta, t \cos\vartheta, a) \, dt \sin\vartheta \, d\vartheta \right) . \end{aligned} \tag{7.32}$$

To continue our calculations we need an analytic expression of I. We do not outline the derivation in detail since this would be very intricate. The interested reader is refered to RIEDER, SCHUSTER [103] where every single step of the proof is described. The expression of I is obtained by using the series representation of the hypergeometric function ${}_2\mathrm{F}_1$ and several technical transformations. At last one gets

$$\begin{aligned} \mathrm{I}(\varphi, s, a) = &\frac{1}{\pi^2} \sin\varphi \, \mathrm{q}(a) \\ &\times \begin{cases} 2\nu(\nu + 1) \Big(s^2 \, {}_2\mathrm{F}_1(1, 1 - \nu; \tfrac{3}{2}; s^2) - 1/(2\nu - 1) \Big), & |s| < 1 , \\ \tfrac{1}{2} \Big(s^{-2} \, {}_2\mathrm{F}_1(1, \tfrac{3}{2}; \nu + 2; s^{-2}) - (2\nu + 2)/(2\nu - 1) \Big), & |s| \geq 1 . \end{cases} \end{aligned} \tag{7.33}$$

It remains to evaluate the double integration over I which occurs on the right-hand side of (7.32). The hypergeometric function ${}_2\mathrm{F}_1$ is given as

$$ {}_2\mathrm{F}_1(u,v;w;z) = \sum_{k=0}^{\infty} \frac{(u)_k\,(v)_k}{(w)_k\,k!}\,z^k\,, $$

where the *Pochhammer symbols* are defined via $(u)_k = u\cdot(u+1)\cdot\ldots\cdot(u+k-1)$ when $k>0$ and $(u)_k = 1$ else. A simple calculation proves

$$ \begin{aligned} \int_0^z t^2\,{}_2\mathrm{F}_1(1,1-\nu;\tfrac{3}{2};t^2)\,\mathrm{d}t &= \sum_{k=0}^{\nu-1} \frac{(1-\nu)_k}{(3/2)_k} \int_0^z t^{2\,k+2}\,\mathrm{d}t \\ &= \sum_{k=0}^{\nu-1} \frac{(1-\nu)_k}{(3/2)_k\,(2\,k+3)}\,z^{2\,k+3}\,. \end{aligned} $$

This identity together with (7.33) leads to

$$ \begin{aligned} \int_{-1}^{1} \mathrm{I}(\vartheta, t\,\cos\vartheta, a)\,\mathrm{d}t &= \frac{4}{\pi^2}\,\nu\,(\nu+1)\,\sin\vartheta\,\mathrm{q}(a) \\ &\quad\times\Big(\sum_{k=0}^{\nu-1} \frac{(1-\nu)_k}{(3/2)_k\,(2\,k+3)}\,(\cos\vartheta)^{2\,k+2} - (2\,\nu-1)^{-1}\Big)\,. \end{aligned} $$

By applying an integration by parts and formula (3.631.17) from GRADSHTEYN, RIZHIK [32] we arrive at

$$ \int_0^{2\pi}\int_{-1}^{1} \sin^2\vartheta\,(\cos\vartheta)^{2\,k+2}\,\mathrm{d}\vartheta = \frac{\pi}{(2\,k+3)\,2^{2k+3}} \binom{2\,k+4}{k+2}\,, $$

which finally yields

$$ \begin{aligned} &\int_0^{2\pi}\int_{-1}^{1} \mathrm{I}(\vartheta, t\,\cos\vartheta, a)\,\mathrm{d}t\,\sin\vartheta\,\mathrm{d}\vartheta = \\ &\qquad \frac{4}{\pi}\,\nu\,(\nu+1)\,\mathrm{q}(a)\left(\sum_{k=0}^{\nu-1} \frac{(1-\nu)_k\binom{2k+4}{k+2}}{(3/2)_k\,(2\,k+3)^2\,2^{2k+3}} - \frac{1}{2\,\nu-1}\right). \end{aligned} \tag{7.34} $$

Putting together (7.30), (7.32), (7.33) and (7.34) we obtain as representation for the unique solution v^1 of (7.24) in $\overline{\mathsf{R}(\mathbf{D}_1)}$:

$$ v^1(\varphi, s, a) = \frac{1}{\pi^2}\,\sin\varphi\,\mathrm{q}(a) \tag{7.35} $$

$$ \times\begin{cases} 2\,\nu\,(\nu+1)\left(s^2\,{}_2\mathrm{F}_1(1,1-\nu;\tfrac{3}{2};s^2) - \tfrac{1}{2\nu-1}\right) - c_\nu\,, & |s|<1\,,\\ \tfrac{1}{2}\left(s^{-2}\,{}_2\mathrm{F}_1(1,\tfrac{3}{2};\nu+2;s^{-2}) - \tfrac{2\nu+2}{2\nu-1}\right) - c_\nu\,, & |s|\ge 1\,, \end{cases} $$

where the constants c_ν are defined by

$$c_\nu = 2\,\nu\,(\nu+1)\left(\sum_{k=0}^{\nu-1} \frac{(1-\nu)_k \binom{2k+4}{k+2}}{(3/2)_k\,(2\,k+3)^2\,2^{2k+3}} - \frac{1}{2\,\nu-1}\right) + \pi\,\frac{2^\nu\,\nu!}{(2\,\nu+1)!!}\,.$$

By the same arguments we get $\upsilon^2 = \upsilon^1$ and υ^3 emerges from υ^1 replacing $\sin\varphi$ in (7.35) by $\cos\varphi$. Figures 7.1, 7.2 show graphic illustrations of the kernel υ^1. Note that the plot in figure 7.1 is very similar to the well-known filter of SHEPP, LOGAN from [117], when the graph is reflected about the x-axis.

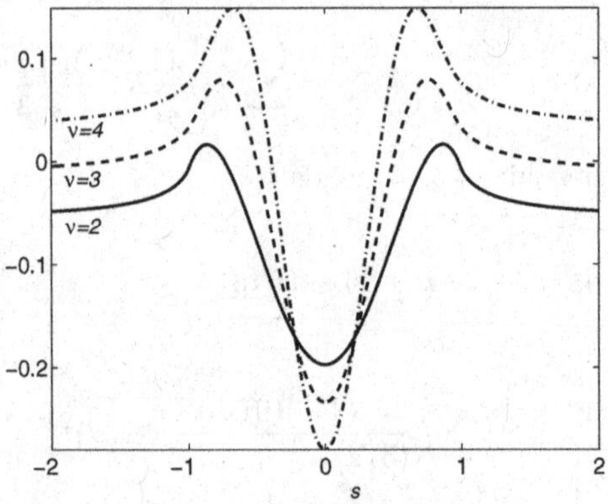

Fig. 7.1. One-dimensional cross section of the reconstruction kernel $\upsilon^1(\pi/2, s, 0)$ (7.35) for $\nu = 2$ (solid curve), $\nu = 3$ (dashed curve) and $\nu = 4$ (dashed-dotted curve). The kernels shape emphasizes that a derivative with respect to s is involved, see equation (7.26).

Remark 7.7. In analogy to the three-dimensional Doppler transform (6.5), (6.6), the *two-dimensional Doppler transform* $\mathbf{D} : L^2(\Omega^2, \mathbb{R}^2) \to L^2(Z)$, $Z = [0, 2\pi] \times [-1, 1]$, is defined by

$$\mathbf{Df}(\varphi, s) = \int_{L(\varphi,s)\cap\Omega^2} \langle \omega^\perp(\varphi), \mathbf{f}(x)\rangle \, \mathrm{d}\ell(x)\,, \tag{7.36}$$

where $\omega^\perp(\varphi) = (-\sin\varphi, \cos\varphi) \in S^1$ is the vector of direction of $L(\varphi, s)$. The transform (7.36) appears as mathematical model when we consider only a two-dimensional cross section through the object. Reconstruction kernels for (7.36) associated with the mollifier $e(x_1, x_2) = \mathrm{p}(x_1, x_2)$ with p from (7.16) are obtained from (7.35) by simply setting $\mathrm{q}(a) \equiv 1$. The method of (semi-discrete) approximate inverse can be formulated according to the three-dimensional case and is outlined in more details in Section 9.2.2. Numerical results are shown in Section 7.3. A measurement device for two-dimensional Doppler tomography was constructed at the Lund Institute of Technology (Sweden) and is described in JANSSON ET AL. [47].

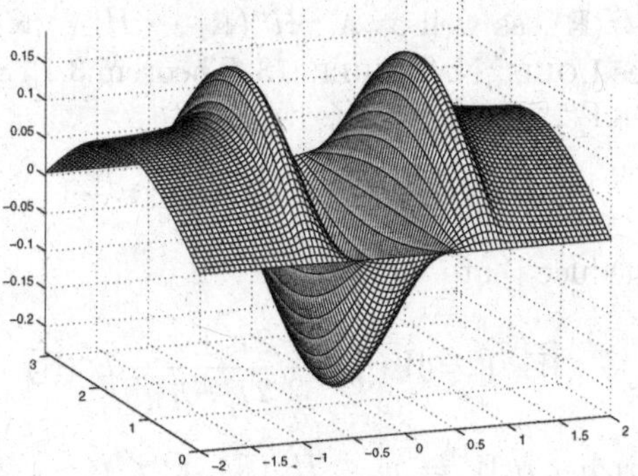

Fig. 7.2. Two-dimensional plot of the reconstruction kernel $v^1(\varphi, s, 0)$ (7.35) for $\nu = 2$, $s \in [-2, 2]$ and $\varphi \in [0, \pi]$. On clearly sees the differentiation in s-direction and the sin-curve in φ-direction, compare representation (7.35).

In view of the convergence theorem 3.6, it is important to have Sobolev norm estimates for the kernels v^j.

Lemma 7.8. *Let e^j be the mollifier according to* (7.15) *and v^j be the corresponding reconstruction kernel, i.e. the unique solution of* (7.24) *in* $\overline{\mathsf{R}(\mathbf{D}_j)}$. *Then v^j can be decomposed to*

$$v^j(\varphi, s, a) = v_1^j(\varphi, s, a) + v_2^j(\varphi, a),$$

where

$$v_1^j \in H^\kappa((0, 2\pi) \times \mathbb{R}) \hat{\otimes} H_0^t(-1, 1) \quad \textit{for } \kappa < \nu,\ t < \nu + 1/2$$

and

$$v_2^j \in H^\kappa(0, 2\pi) \hat{\otimes} H_0^t(-1, 1) \quad \textit{for } \kappa \geq 0,\ t < \nu + 1/2.$$

Proof. Again we only consider the case $j = 1$. The decomposition of v^1 essentially relies on representation (7.30). Function I may be written as

$$\mathrm{I}(\varphi, s, a) = \varrho \sin\varphi\, \mathrm{q}(a)\, W(s)$$

with a constant ϱ and

$$W(s) := \int_{-1}^{s} w(t)\, dt, \qquad w(t) = \int_0^\infty \sigma^{1-\nu}\, \mathrm{J}_{\nu+1}(\sigma)\, \sin(t\,\sigma)\, d\sigma,$$

what becomes obvious when we have a closer look at (7.29) and (7.31). From (7.27) and (7.28) it follows that

$$w(t)\, \sin\varphi = c\,\Lambda\,\mathbf{R}\,\partial_{x_2}\mathrm{p}(\varphi, t)$$

with an appropriate constant c. We further have that $\mathbf{R} : H_0^\kappa(\Omega^2) \to L^2(0, 2\pi)\hat{\otimes} H^{\kappa+1/2}(\mathbb{R})$, as well as $\Lambda : H^\kappa(\mathbb{R}) \to H^{\kappa-1}(\mathbb{R})$ are continuous for any real $\kappa \geq 0$, see LOUIS, NATTERER [73, Theorem 3.1]. Since p is in $H_0^\beta(\Omega^2)$ whenever $\beta < \nu + 1/2$, we have

$$w \in H^\kappa(\mathbb{R}) \quad \text{for all} \quad \kappa < \nu - 1\,.$$

From (7.33) we deduce that

$$\widetilde{W}(\cdot) := W(\cdot) + \frac{1}{2}\frac{2\,\nu + 2}{2\,\nu - 1} \in L^2(\mathbb{R})$$

is valid. Summarizing $\partial_s \widetilde{W} = w \in H^\kappa(\mathbb{R})$, $\kappa < \nu - 1$ and $\widetilde{W} \in L^2(\mathbb{R})$ we conclude that $\widetilde{W} \in H^\kappa(\mathbb{R})$ for any real $\kappa < \nu$. We define

$$v_1^1 = \varrho \sin\varphi\, \mathrm{q}(a)\, \widetilde{W} \quad \text{and} \quad v_2^1 = h(\varphi, a) - \frac{\varrho}{2}\frac{2\,\nu + 2}{2\,\nu - 1} \sin\varphi\, \mathrm{q}(a)\,.$$

Since $h(\varphi, a) = \tilde{c} \sin\varphi\, \mathrm{q}(a)$ for a constant $\tilde{c}$, see (7.32), (7.33), the proof is complete. □

Remark 7.9. In NATTERER, WÜBBELING [84, Theorem 2.27] we find the inversion formula

$$\mathbf{f} = \frac{m-1}{2\pi\,|S^{m-2}|}\,\Lambda^{-\alpha}\,\mathbf{D}_\mathrm{p}^*\,\Lambda^{\alpha-1}\,\mathbf{D}_\mathrm{p}\mathbf{f}\,, \qquad \alpha < m\,, \tag{7.37}$$

where

$$\mathbf{D}_\mathrm{p}\mathbf{f}(\omega, x) = \int_{-\infty}^{\infty} \omega \cdot \mathbf{f}(x + t\,\omega)\,\mathrm{d}t\,, \qquad x \in \{\omega^\perp\}\,,$$

which is valid for solenoidal $\mathbf{f} \in \mathcal{S}(\mathbb{R}^m, \mathbb{R}^m)$. In case that we consider *all* lines in (6.3), $\mathbf{D}_\mathrm{p}$ and the Doppler transform $\mathbf{D}$ coincide. In that case reconstruction kernels can also be computed by means of (7.37) for solenoidal mollifiers $e_{d,k}^j\,\delta_j$. Our specific mollifier (7.15) is not solenoidal and the lines of integration in (6.5), (6.6) are restricted due to the special measurement geometry. That is why we have considered the normal equations (7.24).

7.3 The method of approximate inverse for $\Psi_{p,q,r}\,\mathbf{D}$

Suppose we have a solution v^j of (7.23) available. According to Corollary 7.3 we put

$$v_{d,k}^j = \mathcal{G}_j^{d,k} v^j\,, \quad d \geq 1\,, \quad k \in \mathbb{Z}^3\,, \quad \|k\| \leq d - 1\,.$$

Due to the notations and investigations made in Section 7.1, we are now able to state the semi-discrete approximate inverse $\widetilde{\mathbf{D}}_{j,n,d} : \mathbb{R}^n \to L^2(\Omega^3)$ associated with the mapping $\Psi_{p,q,r}\,\mathbf{D}_j$ by

$$\widetilde{\mathbf{D}}_{j,n,d}v(x) := \sum_{\substack{k\in\mathbb{Z}^3\\ \|k\|\le d-1}} \langle v, G_{p,q,r}\,\Psi_{p,q,r}\,\mathcal{G}_j^{d,k}v^j\rangle_{\mathbb{R}^n}\, B_{d,k}(x)\,, \quad x\in\Omega^3\,. \quad (7.38)$$

The definition is due to the abstract setting (3.17). Here, $\{B_{d,k}\}$ denotes the Riesz system (7.13) and $G_{p,q,r}\in\mathbb{R}^{n\times n}$ is the Gramian matrix corresponding to the splines (7.1)

$$(G_{p,q,r})_{(l_\mu,i_\mu,k_\mu),(l_\lambda,i_\lambda,k_\lambda)} = \langle b^{(0)}_{p,l_\mu}\otimes b^{(0)}_{q,i_\mu}\otimes b^{(0)}_{r,k_\mu}, b^{(0)}_{p,l_\lambda}\otimes b^{(0)}_{q,i_\lambda}\otimes b^{(0)}_{r,k_\lambda}\rangle_{L^2(Q)}\,.$$

A simple calculation shows that $G_{p,q,r}$ is a multiple of the identity matrix

$$G_{p,q,r} = \frac{2\,\pi}{p\,q\,r}\,I_{n,n}\,.$$

Putting the measured data $v = \Psi_{p,q,r}\,\mathbf{D}_j\mathbf{f}$ in (7.38) for a vector field $\mathbf{f}\in\mathcal{X}^{\alpha,\beta}$, $\alpha,\beta>1/2$, then the inner products in (7.38) compute as

$$\langle\Psi_{p,q,r}\,\mathbf{D}_j\mathbf{f}, G_{p,q,r}\,\Psi_{p,q,r}\,\mathcal{G}_j^{d,k}v^j\rangle_{\mathbb{R}^n} =$$

$$\frac{2\,\pi\,d^3}{p\,q\,r}\sum_{m=-r}^{r-1}\sum_{l=0}^{p-1}\sum_{i=-q}^{q-1}\mathbf{D}_j\mathbf{f}(\varphi_l,s_i,a_m)v^j\Big(\varphi_l, d\,s_i - \langle\mathcal{P}_j^* k,\omega(\varphi_l)\rangle, d\,a_m - \langle k,w_j\rangle\Big).$$

Thus the scalar products in (7.38) can be evaluated very efficiently by a method of *filtered backprojection* type, a distinguished method well-known in computerized tomography. We refer the reader to NATTERER [80, Chapter V.1] for a comprehensive study of the filtered backprojection algorithm. If we use the particular representation (7.35) and the fact that $\mathrm{q}(d\,a - \langle k,w_j\rangle)\approx \mathrm{q}(d^{-1}\langle k,w_j\rangle)$ for large d, then we can cancel the outer summation saving one order of time complexity. The details of that efficiency increase as well as hints for the implementation of the method are outlined in [110].

The semi-discrete approximate inverse $\widetilde{\mathbf{D}}_{n,d}:\mathbb{R}^{3n}\to L^2(\Omega^3,\mathbb{R}^3)$ of $\Psi_{p,q,r}\,\mathbf{D}$ is finally defined as

$$(\widetilde{\mathbf{D}}_{n,d}v)_j(x) := \widetilde{\mathbf{D}}_{j,n,d}v_j(x)\,,\quad v=(v_1,v_2,v_3)\,,\quad v_j\in\mathbb{R}^n\,,\quad x\in\Omega^3\,. \quad (7.39)$$

Figures 7.3, 7.4 show reconstructions of solenoidal vector fields. In [110] the author emphasizes that it is also possible to recover the curl of the field $\nabla\times\mathbf{f}$ by a simple change of the reconstruction kernel. This is an additional feature of this method. Figure 7.4 displays a reconstruction of the curl of a solenoidal field.

Chapter 8 aims to prove the limits

$$\lim_{\substack{n\to\infty\\ d\to\infty}}\widetilde{\mathbf{D}}_{j,n,d}\,\Psi_{p,q,r}\,\mathbf{D}_j\mathbf{f} = (\mathcal{P}_{\mathsf{N}(\mathbf{D}_j)^\perp}\mathbf{f})_j\,,\quad j=1,2,3 \quad (7.40)$$

and

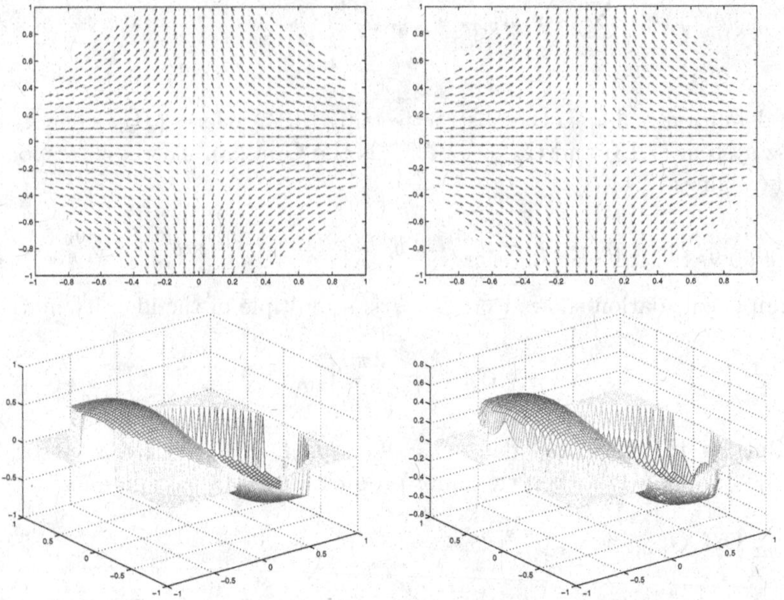

Fig. 7.3. Reconstruction of $\mathbf{f}(x_1, x_2) = 2\exp(-x_1^2 - x_2^2)(-x_2, x_1)$ from discrete Doppler data. Left column: Original vector field $\mathbf{f}$ and first component $\mathbf{f}_1$. Right column: Reconstruction of $\mathbf{f}$ and of its first component.

'The picture is taken from T. Schuster, *Defect correction in vector field tomography: detecting the potential part using BEM and implementation of the method*, Inverse Problems, 21 (2005), pp. 75–91. Copyright ©2005 IOP Publishing Limited. Reprinted with permission.'

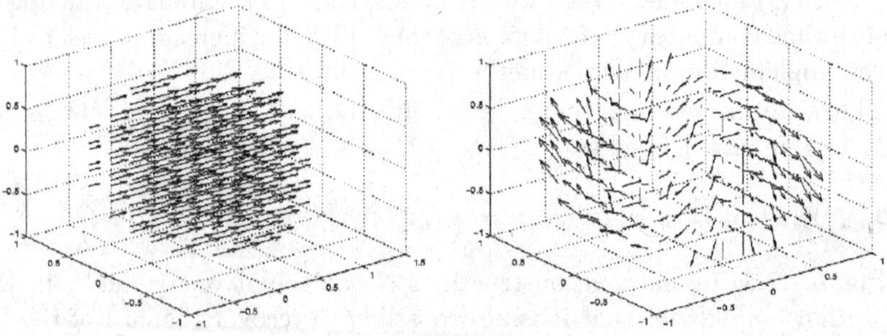

Fig. 7.4. Reconstruction of the vector field $\mathbf{f}(x_1, x_2, x_3) = (1 - x_2^2 - x_3^2, 0, 0)$ which describes a horizontal, solenoidal flow through a cylinder centered about the x_3-axis using the method of approximate inverse (left picture). The right picture shows a reconstruction of the curl of the field $\nabla \times \mathbf{f}$. The curl $\nabla \times \mathbf{f}(x_1, x_2, x_3) = (0, -2x_3, 2x_2)$ consists of a vortex which is clearly visible in the picture to the right.

'The picture is taken from T. Schuster, *An efficient mollifier method for three-dimensional vector tomography: convergence analysis and implementation*, Inverse Problems, 17 (2001), pp. 739–766. Copyright ©2001 IOP Publishing Limited. Reprinted with permission.'

$$\lim_{\substack{n\to\infty\\ d\to\infty}} \widetilde{\mathbf{D}}_{n,d}\,\Psi_{p,q,r}\,\mathbf{D}\mathbf{f} = \mathcal{P}f\,, \quad (\mathcal{P}\mathbf{f})_j := (\mathcal{P}_{\mathsf{N}(\mathbf{D}_j)^\perp}\mathbf{f})_j\,, \quad j = 1,2,3\,. \tag{7.41}$$

Note that $\mathcal{P} : L^2(\Omega^3, \mathbb{R}^3) \to L^2(\Omega^3, \mathbb{R}^3)$ actually is a projection which is not orthogonal. The convergences (7.40) and (7.41) are the best we can accomplish since $\mathbf{D}_j$ and $\mathbf{D}$ are not one-to-one. However, we have explicit representations for the null spaces $\mathsf{N}(\mathbf{D}_j)$, $\mathsf{N}(\mathbf{D})$, see Remark 6.4.

8

Convergence and stability

This chapter is concerned with the application of convergence Theorem 3.6 to the inversion method (7.39). To this end, we will need not only the estimates (7.3), (7.5) and (7.19) of the operators $\Pi_{p,q,r}$ and $\mathbf{E}_d$, respectively, but we have also to investigate the behavior of the kernels $v^j_{d,k}$ in Sobolev norms. The following lemma is a first step into this direction. Much of this chapter is subject of the article [103] of RIEDER AND SCHUSTER.

Lemma 8.1. *We put* $Z = (0, 2\pi) \times (-1, 1)$, $Z_j^{d,k} = \{(\varphi, d\,s - \langle \mathcal{P}_j^* k, \omega(\varphi)\rangle) : (\varphi, s) \in Z\}$ *and* $I_j^{d,k} = \{d\,a - \langle k, w_j\rangle : a \in (-1,1)\}$. *If* $\|k\| \le d$, *then there exists a constant* $C_{\mathcal{G}} > 0$ *with*

$$\|\mathcal{G}_j^{d,k} g\|_{H^\kappa(Z)\hat{\otimes} H^\tau(-1,1)} \le C_{\mathcal{G}}\, d^{\kappa+\tau+2}\, \|g\|_{H^\kappa(Z_j^{d,k})\hat{\otimes} H^\tau(I_j^{d,k})}, \tag{8.1}$$

whenever the right-hand side is bounded for $\kappa, \tau \ge 0$.

Proof. Let $\|k\| \le d$. We have $\mathcal{G}_j^{d,k} = T_j^{d,k} \otimes K_j^{d,k}$, where

$$T_j^{d,k} v(\varphi, s) := d^2\, v(\varphi, d\,s - \langle \mathcal{P}_j^* k, \omega(\varphi)\rangle) \tag{8.2}$$

and

$$K_j^{d,k} u(a) := d\, u(d\,a - \langle k, w_j\rangle)\,.$$

First, we prove

$$\|T_j^{d,k} v\|_{H^\kappa(Z)} \le C_T\, d^{\kappa+3/2}\, \|v\|_{H^\kappa(Z_j^{d,k})} \tag{8.3}$$

for a constant $C_T > 0$.

The transform $\Phi_j(\varphi, s) := (\varphi, d\,s - \langle \mathcal{P}_j^* k, \omega(\varphi)\rangle)$ represents a $\mathcal{C}^\infty$-diffeomorphism between Z and $Z_j^{d,k}$, such that $T_j^{d,k} v = d^2\, v \circ \Phi_j$. The Jacobian of Φ_j satisfies $\det J\Phi_j(\varphi, s) = d$. We prove by complete induction that

$$|v \circ \Phi_j|_{H^\kappa(Z)} \le \tilde{c}_T\, d^{\kappa-1/2}\, |v|_{H^\kappa(Z_j^{d,k})} \tag{8.4}$$

for a $\tilde{c}_T > 0$. Here, $|v|^2_{H^\kappa(Z)} = \sum_{|\alpha|=\kappa} \|D^\alpha v\|_{L^2(Z)}$, $\alpha \in \mathbb{N}_0^2$, denotes the Sobolev seminorm on $H^\kappa(Z)$.
For $\kappa = 0$ estimate (8.4) is obtained by simple integral transformations. Suppose that (8.4) is valid for an arbitrary $\kappa \in \mathbb{N}$. We have

$$|v \circ \Phi_j|^2_{H^{\kappa+1}(Z)} \le |D^{(1,0)}(v \circ \Phi_j)|^2_{H^\kappa(Z)} + |D^{(0,1)}(v \circ \Phi_j)|^2_{H^\kappa(Z)}\,.$$

A closer look to the different differential operators yields

$$\begin{aligned} D^{(0,1)}(v \circ \Phi_j) &= d\,(D^{(0,1)}v) \circ \Phi_j\,, \\ D^{(1,0)}(v \circ \Phi_j) &= \langle \mathcal{P}_j^* k, \omega^\perp(\varphi)\rangle\,(D^{(0,1)}v) \circ \Phi_j + (D^{(1,0)}v) \circ \Phi_j\,. \end{aligned}$$

From the induction hypothesis, we deduce

$$\begin{aligned} |D^{(0,1)}(v \circ \Phi_j)|^2_{H^\kappa(Z)} &= d^2\,|(D^{(0,1)}v) \circ \Phi_j|^2_{H^\kappa(Z)} \\ &\le \tilde{c}_T\, d^{2\kappa+1}\,|D^{(0,1)}v|^2_{H^\kappa(Z)} \le \tilde{c}_{T,1}\, d^{2\kappa+1}\,|v|^2_{H^{\kappa+1}(Z)}\,. \end{aligned}$$

If we take into account that $|\langle \mathcal{P}_j^* k, \omega^\perp(\varphi)\rangle| \le d$, then the same argumentation gives

$$|D^{(1,0)}(v \circ \Phi_j)|^2_{H^\kappa(Z)} \le \tilde{c}_{T,2}\, d^{2\kappa+1}\,|g|^2_{H^{\kappa+1}(Z_j^{d,k})}$$

with an additional constant $\tilde{c}_{T,2} > 0$, which completes the proof of (8.4). We finally get

$$\|T_j^{d,k} v\|^2_{H^\kappa(Z)} = d^4\,\|v \circ \Phi_j\|^2_{H^\kappa(Z)} = d^4 \sum_{i=0}^{\kappa} |v \circ \Phi_j|^2_{H^i(Z)} \le \tilde{c}_T\, d^{2\kappa+3}\,\|v\|^2_{H^\kappa(Z_j^{d,k})}$$

yielding (8.3) for $\kappa \in \mathbb{N}_0$ with $C_T = \sqrt{\tilde{c}_T}$. Estimate (8.3) is obtained for any real $\kappa \ge 0$ by an application of the interpolation inequality for Sobolev spaces, see e.g. LIONS, MAGENES [64, Chapter 5.1].
In the same way

$$\|K_j^{d,k} u\|_{H^\tau(-1,1)} \le C_K\, d^{\tau+1/2}\,\|u\|_{H^\tau(I_j^{d,k})}$$

is verified for a constant $C_K > 0$. Using the tensor product structure of $\mathcal{G}_j^{d,k}$ finishes the proof. □

From (8.1) we deduce the necessary Sobolev norm estimates of $\|v_j^{d,k}\|_{\mathcal{Y}^{\alpha,\beta}}$.

Corollary 8.2. *Let the mollifiers e^j, $j = 1,2,3$ be defined according to (7.15), (7.16), (7.17). If $\nu > \max\{\alpha + 1/2, \beta - 1/2\}$ and $d \ge 1$, then there exists a constant $C_v > 0$ with*

$$\|v^j_{d,k}\|_{\mathcal{Y}^{\alpha,\beta}} \le C_v\, d^{\alpha+\beta+5/2}\,, \qquad j = 1,2,3\,. \tag{8.5}$$

The boundedness is uniform in k for $\|k\| \le d-1$.

Proof. Due to Lemma 7.8 υ^j allows for the splitting

$$\upsilon^j(\varphi, s, a) = \upsilon_1^j(\varphi, s, a) + \upsilon_2^j(\varphi, a),$$

where $\upsilon_1^j \in H^{\alpha+1/2}((0, 2\pi) \times \mathbb{R})\hat{\otimes} H^\beta(\mathbb{R})$ and $\upsilon_2^j \in H^{\alpha+1/2}(0, 2\pi)\hat{\otimes} H^\beta(\mathbb{R})$. Applying (8.1) leads to

$$\begin{aligned}\|\upsilon_{d,k}^j\|_{\mathcal{Y}^{\alpha,\beta}} &\le \|\mathcal{G}_j^{d,k}\upsilon_1^j\|_{\mathcal{Y}^{\alpha,\beta}} + \|\mathcal{G}_j^{d,k}\upsilon_2^j\|_{\mathcal{Y}^{\alpha,\beta}} \\ &\le C_{\mathcal{G}}\Big(d^{\alpha+\beta+5/2}\,\|\upsilon_1^j\|_{H^{\alpha+1/2}((0,2\pi)\times\mathbb{R})\hat{\otimes} H^\beta(\mathbb{R})} \\ &\quad + d^{\beta+5/2}\,\|\upsilon_2^j\|_{H^{\alpha+1/2}(0,2\pi)\hat{\otimes} H^\beta(\mathbb{R})}\Big) \le C_\upsilon\, d^{\alpha+\beta+5/2},\end{aligned}$$

where we have defined

$$C_\upsilon := C_{\mathcal{G}}\left(\|\upsilon_1^j\|_{H^{\alpha+1/2}((0,2\pi)\times\mathbb{R})\hat{\otimes} H^\beta(\mathbb{R})} + \|\upsilon_2^j\|_{H^{\alpha+1/2}(0,2\pi)\hat{\otimes} H^\beta(\mathbb{R})}\right).$$

□

We have now all ingredients together to prove the convergence (7.40).

Theorem 8.3. *Let $\alpha, \beta > 1/2$ and $\mathbf{f} \in \mathcal{X}^{\alpha,\beta}$ having support in Ω^3. Suppose the mollifiers e^j, $j = 1, 2, 3$, to be defined by (7.15), (7.16) and (7.17), where $\nu > \max\{\alpha+1/2, \beta-1/2\}$. We further denote by $\tilde{d} = \tilde{d}(\mathbf{f})$ the smallest positive integer such that supp $\mathbf{f} \subset B_{1-1/\tilde{d}}(0)$*[1]*. If $d \ge \tilde{d}$, then we have for $j = 1, 2, 3$ the estimate*

$$\begin{aligned}\|\widetilde{\mathbf{D}}_{j,n,d}\,\Psi_{p,q,r}\,\mathbf{D}_j\mathbf{f} - (\mathcal{P}_{\mathsf{N}(\mathbf{D}_j)^\perp}\mathbf{f})_j\|_{L^2(\Omega^3)} \\ \le C\left(d^{-\min\{2,\alpha\}} + d^{-\min\{2,\beta\}} + \rho\, d^{\alpha+\beta+5/2}\right)\end{aligned} \tag{8.6}$$

with a constant $C > 0$.

Proof. Because of $\langle \mathbf{f}, e_{d,k}^j\,\delta_j\rangle_{L^2(\Omega^3,\mathbb{R}^3)} = 0$ for $d \ge \tilde{d}$ and $\|k\| \ge d$, the mollifier operator $(\mathbf{E}_d)_j$ has the representation

$$(\mathbf{E}_d)_j\mathbf{f}(x) = \sum_{\substack{k\in\mathbb{Z}^3\\ \|k\|\le d-1}} \langle \mathbf{f}, e_{d,k}^j\,\delta_j\rangle_{L^2(\Omega^3,\mathbb{R}^3)}\, b_{d,k}(x).$$

As a consequence of Theorem 3.6 and Corollary 8.2, we obtain

$$\begin{aligned}\|\widetilde{\mathbf{D}}_{j,n,d}\,\Psi_{p,q,r}\,\mathbf{D}_j\mathbf{f} - (\mathcal{P}_{\mathsf{N}(\mathbf{D}_j)^\perp}\mathbf{f})_j\|_{L^2(\Omega^3)} \\ \le C\,\rho\, d^{\alpha+\beta+5/2}\,\|\mathbf{f}\|_{\mathcal{X}^{\alpha,\beta}} + \left\|\Big((I - \mathbf{E}_d)\,\mathcal{P}_{\mathsf{N}(\mathbf{D}_j)^\perp}\mathbf{f}\Big)_j\right\|_{L^2(\Omega^3)}.\end{aligned}$$

Since $f \in \mathcal{X}^{\alpha,\beta}$ we either have $\mathcal{P}_{\mathsf{N}(\mathbf{D}_j)^\perp}\mathbf{f} \in \mathcal{X}^{\alpha,\beta}$ in which case (8.6) follows directly from (7.19), or $\mathcal{P}_{\mathsf{N}(\mathbf{D}_j)}\mathbf{f} \in \mathcal{X}^{\alpha,\beta}$. In the latter case, we use

[1] $B_R(z)$ is the open ball with radius $R > 0$ centered about z, e.g. $B_1(0) = \Omega^3$.

$$\left\|\left((I-\mathbf{E}_d)\,\mathcal{P}_{\mathrm{N}(\mathbf{D}_j)^\perp}\mathbf{f}\right)_j\right\|_{L^2(\Omega^3)} = \left\|\left((I-\mathbf{E}_d)\,(I-\mathcal{P}_{\mathrm{N}(\mathbf{D}_j)})\mathbf{f}\right)_j\right\|_{L^2(\Omega^3)}$$

and prove (8.6) after an application of the triangle inequality together with (7.19). □

The estimate (8.6) helps us to find a relation between the data sampling which is characterized by the discretizations p, q and r and the regularization parameter d, such that we are able to prove convergence with rates.

Corollary 8.4. *Adopt all assumptions made in Theorem 8.3. Let further sequences $\{p_\mu\}_{\mu\in\mathbb{N}} \subset \mathbb{N}$, $\{q_\mu\}_{\mu\in\mathbb{N}} \subset \mathbb{N}$, $\{r_\mu\}_{\mu\in\mathbb{N}} \subset \mathbb{N}$ and a sequence $\{d_\mu\}_{\mu\in\mathbb{N}} \subset \mathbb{R}^+$ be given in such a way, that all the sequences diverge to infinity as $\mu \to \infty$ and that the relations*

$$\lim_{\mu\to\infty} d_\mu^{\alpha+\beta+5/2}/\min\{p_\mu, q_\mu\} = \lim_{\mu\to\infty} d_\mu^{\alpha+\beta+5/2}/r_\mu^{\min\{\beta,1\}} = 0$$

are valid. Then we have the convergences

$$\lim_{\mu\to\infty} \|\widetilde{\mathbf{D}}_{j,n_\mu,d_\mu}\,\Psi_{p_\mu,q_\mu,r_\mu}\,\mathbf{D}_j\mathbf{f} - (\mathcal{P}_{\mathrm{N}(\mathbf{D}_j)^\perp}\mathbf{f})_j\|_{L^2(\Omega^3)} = 0\,, \quad j = 1,2,3\,, \tag{8.7}$$

and

$$\lim_{\mu\to\infty} \|\widetilde{\mathbf{D}}_{n_\mu,d_\mu}\,\Psi_{p_\mu,q_\mu,r_\mu}\,\mathbf{D}\mathbf{f} - \mathcal{P}\mathbf{f}\|_{L^2(\Omega^3,\mathbb{R}^3)} = 0\,.$$

If the quantities p, q, r and d further allow for a coupling[2] *$p \simeq q$, $r \simeq q^{1/\min\{\beta,1\}}$ und $d \simeq q^\lambda$, where*

$$\lambda = \lambda(\alpha,\beta) = \frac{1}{\alpha+\beta+5/2+\min\{2,\alpha,\beta\}}\,, \tag{8.8}$$

then for $q \to \infty$ we get the convergences with rates

$$\|\widetilde{\mathbf{D}}_{j,n,d}\,\Psi_{p,q,r}\,\mathbf{D}_j\mathbf{f} - (\mathcal{P}_{\mathrm{N}(\mathbf{D}_j)^\perp}\mathbf{f})_j\|_{L^2(\Omega^3)} \le C_1\,q^{-\lambda\,\min\{2,\alpha,\beta\}}$$

as well as

$$\|\widetilde{\mathbf{D}}_{n,d}\,\Psi_{p,q,r}\,\mathbf{D}\mathbf{f} - \mathcal{P}\mathbf{f}\|_{L^2(\Omega^3,\mathbb{R}^3)} \le C_2\,q^{-\lambda\,\min\{2,\alpha,\beta\}}$$

with constants $C_1, C_2 > 0$.

Proof. At first, we prove the convergence statements for $\widetilde{\mathbf{D}}_{j,n,d}$. Convergence (8.7) is an immediate consequence from (8.6) and the requirements to $\{p_\mu\}$, $\{q_\mu\}$, $\{r_\mu\}$ and $\{d_\mu\}$. By the assumptions made for p, q and r, we learn from (7.4) that $\rho \simeq q^{-1}$. This implies

[2] By $A \simeq B$ we mean the existence of generic constants $c_1, c_2 > 0$ validating $c_1\,A \le B \le c_2\,A$.

$$\|\widetilde{\mathbf{D}}_{j,n,d}\, \Psi_{p,q,r}\, \mathbf{D}_j \mathbf{f} - (\mathcal{P}_{\mathsf{N}(\mathbf{D}_j)^\perp}\mathbf{f})_j\|_{L^2(\Omega^3)} \\ \le C\left(q^{-\lambda\,\min\{2,\alpha\}} + q^{-\lambda\,\min\{2,\beta\}} + q^{-1+\lambda\,(\alpha+\beta+5/2)}\right).$$

Because of

$$-1 + \lambda\,(\alpha+\beta+5/2) = -\lambda\,\min\{2,\alpha,\beta\}$$

this gives the convergence rate for $\widetilde{\mathbf{D}}_{j,n,d}$ with $C_1 = C$. The corresponding results for $\widetilde{\mathbf{D}}_{n,d}$ follow from those for $\widetilde{\mathbf{D}}_{j,n,d}$ taking into account that the norm on $L^2(\Omega^3,\mathbb{R}^3)$ is given as $\|\mathbf{f}\|^2_{L^2(\Omega^3,\mathbb{R}^3)} = \sum_{j=1}^3 \|\mathbf{f}_j\|^2_{L^2(\Omega^3)}$. □

From Corollary 8.4 it is easy to derive convergence in the 2D case. In contrast to the three-dimensional Doppler transform, we even get exact convergence as $d, n \longrightarrow \infty$ in a suitable manner, if only the field $\mathbf{f}$ is solenoidal. The semi-discrete approximate inverse for the two-dimensional Doppler transform is described in more detail in section 9.2.2.

Corollary 8.5. *Let* $\mathbf{D} : L^2(\Omega^2,\mathbb{R}^2) \to L^2(Z)$ *be the two-dimensional Doppler transform* (7.36), $\Psi_{p,q}$ (6.9) *the corresponding observation operator of the semi-discrete problem and the mollifier be defined by* $\mathrm{e}(x_1,x_2) = \mathrm{p}(x_1,x_2)$ *with* p *as in* (7.16). *Assume further that* $\mathbf{f} \in H^\alpha(\Omega^2,\mathbb{R}^2)$ *for* $\alpha > 1/2$ *and that sequences* $\{p_\mu\}_{\mu\in\mathbb{N}} \subset \mathbb{N}$, $\{q_\mu\}_{\mu\in\mathbb{N}} \subset \mathbb{N}$ *and* $\{d_\mu\}_{\mu\in\mathbb{N}} \subset \mathbb{R}^+$ *are given in such a way that all the sequences diverge to infinity as* $\mu \longrightarrow \infty$ *and that the relation*

$$\lim_{\mu\to\infty} d_\mu^{\alpha+5/2} / \min\{p_\mu, q_\mu\} = 0$$

is valid. Then

$$\lim_{\mu\to\infty} \|\widetilde{\mathbf{D}}_{n_\mu,d_\mu}\, \Psi_{p_\mu,q_\mu}\, \mathbf{D}\mathbf{f} - \mathcal{P}_{\mathsf{N}(\mathbf{D})^\perp}\mathbf{f}\|_{L^2(\Omega^2,\mathbb{R}^2)} = 0\,.$$

Since $\mathbf{f} = \mathcal{P}_{\mathsf{N}(\mathbf{D})^\perp}\mathbf{f}$ *if* $\nabla\cdot\mathbf{f} = 0$ *in* Ω^2, *we have exact convergence for solenoidal fields*

$$\lim_{\mu\to\infty} \|\widetilde{\mathbf{D}}_{n_\mu,d_\mu}\, \Psi_{p_\mu,q_\mu}\, \mathbf{D}\mathbf{f} - \mathbf{f}\|_{L^2(\Omega^2,\mathbb{R}^2)} = 0\,. \tag{8.9}$$

Proof. The convergence can be deduced from Corollary 8.4 with $\beta = 0$. To get the exact convergence (7.41) we only have to show that $\mathbf{f} = \mathcal{P}_{\mathsf{N}(\mathbf{D})^\perp}\mathbf{f}$ for solenoidal fields. To this end, assume $\nabla\cdot\mathbf{f} = 0$ and take $\mathbf{h} \in \mathsf{N}(\mathbf{D})$. We have $\mathbf{h} = \nabla p$ with $p = 0$ at $\partial\Omega^2$, compare Remark 6.4. Hence,

$$\int_{\Omega^2} \mathbf{f}(x)\cdot\mathbf{h}(x)\,\mathrm{d}x = \int_{\partial\Omega^2} (\nu_x\cdot\mathbf{f}(x))\,p(x)\,\mathrm{d}s_x - \int_{\Omega^2} \nabla\cdot\mathbf{f}(x)\,p(x)\,\mathrm{d}x = 0$$

by the Gauss-Ostrogradsky theorem. Here, ν_x denotes the unit outer normal vector of $\partial\Omega^2$. This shows $\mathbf{f} \perp \mathsf{N}(\mathbf{D})$, which completes the proof. □

We conclude this chapter by having a closer look at the regularization properties of the approximate inverses $\widetilde{\mathbf{D}}_{j,n,d}$, $\widetilde{\mathbf{D}}_{n,d}$, respectively, where our investigations rely on the abstract framework set in Theorem 3.10. Compared to the definitions of optimality made in Definition 1.5 we will realize that the presented inversion method represents a regularization which is sub-optimal.

We assume the measured data $\Psi_{p,q,r}\,\mathbf{D}_j\mathbf{f}$ to be contaminated by noise, where the perturbation again is interpretated as error in the observation operator $\Psi_{p,q,r}$. Hence, let $\Psi^\eta_{p,q,r} : \mathcal{Y}^{\alpha,\beta} \to \mathbb{R}^n$ for $\alpha, \beta > 1/2$ be a contamination of $\Psi_{p,q,r}$ with noise level $\eta > 0$

$$\left|(\Psi^\eta_{p,q,r}v)_{l,i,m} - (\Psi_{p,q,r}v)_{l,i,m}\right| \le \eta\,\|v\|_{\mathcal{Y}^{\alpha,\beta}}\,, \qquad v \in \mathcal{Y}^{\alpha,\beta} \tag{8.10}$$

for all $0 \le l \le p-1$, $-q \le i \le q-1$ and $-r \le m \le r-1$. Corollary 8.6 tells how to choose the number of data scanning points and the regularization parameter d subject to the noise level η to yield convergence as $\eta \to 0$, see also Remark 2.5.

Corollary 8.6. *Adopt again all assumptions made in Theorem 8.3. Let further $p \simeq q$, $r \simeq q$ and $d \simeq q^\lambda$ with λ as in (8.8). The noise perturbed observation operator $\Psi^\eta_{p,q,r}$ is supposed to satisfy (8.10). If $q \simeq \eta^{-1}$, then there are constants $\widetilde{C}_1, \widetilde{C}_2 > 0$ validating*

$$\|\widetilde{\mathbf{D}}_{j,n,d}\,\Psi^\eta_{p,q,r}\,\mathbf{D}_j\mathbf{f} - (\mathcal{P}_{\mathsf{N}(\mathbf{D}_j)^\perp}\mathbf{f})_j\|_{L^2(\Omega^3)} \le \widetilde{C}_1\,\eta^{\frac{\min\{2,\alpha,\beta\}}{\alpha+\beta+5/2+\min\{2,\alpha,\beta\}}}$$

as well as

$$\|\widetilde{\mathbf{D}}_{n,d}\,\Psi^\eta_{p,q,r}\,\mathbf{D}\mathbf{f} - \mathcal{P}\mathbf{f}\|_{L^2(\Omega^3,\mathbb{R}^3)} \le \widetilde{C}_2\,\eta^{\frac{\min\{2,\alpha,\beta\}}{\alpha+\beta+5/2+\min\{2,\alpha,\beta\}}}$$

for $\eta \to 0$.

Proof. We split the reconstruction error into data and approximation error which is a usual procedure in the theory of regularization methods,

$$\begin{aligned}\|\widetilde{\mathbf{D}}_{j,n,d}\,\Psi^\eta_{p,q,r}\,\mathbf{D}_j\mathbf{f} - (\mathcal{P}_{\mathsf{N}(\mathbf{D}_j)^\perp}\mathbf{f})_j\|_{L^2(\Omega^3)} \le\\ \|\widetilde{\mathbf{D}}_{j,n,d}\,(\Psi^\eta_{p,q,r} - \Psi_{p,q,r})\,\mathbf{D}_j\mathbf{f}\|_{L^2(\Omega^3)} +\\ \|\widetilde{\mathbf{D}}_{j,n,d}\,\Psi_{p,q,r}\,\mathbf{D}_j\mathbf{f} - (\mathcal{P}_{\mathsf{N}(\mathbf{D}_j)^\perp}\mathbf{f})_j\|_{L^2(\Omega^3)}\,.\end{aligned}$$

From the proofs of Theorem 3.10 and (8.5) we see that the data error can be estimated as

$$\begin{aligned}\|\widetilde{\mathbf{D}}_{j,n,d}\,(\Psi^\eta_{p,q,r} - \Psi_{p,q,r})\,\mathbf{D}_j\mathbf{f}\|_{L^2(\Omega^3)} &\le c\eta\left(d^{-1}\sum_{i=1}^{d}\|v^j_{d,k}\|^2_{\mathcal{Y}^{\alpha,\beta}}\right)^{1/2}\|\mathbf{f}\|_{\mathcal{X}^{\alpha,\beta}}\\ &\le c\,C_v\,\eta\,d^{\alpha+\beta+5/2}\,\|\mathbf{f}\|_{\mathcal{X}^{\alpha,\beta}}\end{aligned}$$

for suitable $c > 0$. The approximation error is obtained from Corollary 8.4,

$$\|\widetilde{\mathbf{D}}_{j,n,d}\,\Psi_{p,q,r}\,\mathbf{D}_j\mathbf{f} - (\mathcal{P}_{\mathsf{N}(\mathbf{D}_j)^\perp}\mathbf{f})_j\|_{L^2(\Omega^3)} \le C_1\,q^{-\lambda\,\min\{2,\alpha,\beta\}}\,.$$

The total error is then given as

$$\begin{aligned}\|\widetilde{\mathbf{D}}_{j,n,d}\,\Psi^{\eta}_{p,q,r}\,\mathbf{D}_j\mathbf{f} - (\mathcal{P}_{\mathsf{N}(\mathbf{D}_j)^\perp}\mathbf{f})_j\|_{L^2(\Omega^3)} &\le \bar{C}_1\Big(\eta\, q^{\lambda\,(\alpha+\beta+5/2)} + q^{-\lambda\,\min\{2,\alpha,\beta\}}\Big)\\ &\le 2\,\bar{C}_1\,\eta^{\frac{\min\{2,\alpha,\beta\}}{\alpha+\beta+5/2+\min\{2,\alpha,\beta\}}}\,,\end{aligned}$$

where $\bar{C}_1 := \max\{c\,C_v\,\|\mathbf{f}\|_{\mathcal{X}^{\alpha,\beta}}, C_1\}$. This yields also the reconstruction error in $\widetilde{\mathbf{D}}_{n,d}\,\Psi^{\eta}_{p,q,r}\,\mathbf{Df}$ after applying the triangle inequality. □

Remark 8.7. In view of Definition 1.5, Corollary 8.6 tells us that the inversion scheme $\widetilde{\mathbf{D}}_{n,d}$ represents a regularization which is sub-optimal. More explicitly, if $1/2 < \alpha = \beta \le 2$, then

$$\|\widetilde{\mathbf{D}}_{n,d}\,\Psi^{\eta}_{p,q,r}\,\mathbf{Df} - \mathcal{P}\mathbf{f}\|_{L^2(\Omega^3,\mathbb{R}^3)} \le \widetilde{C}_2\,\eta^{\frac{\alpha}{3\alpha+5/2}}\,,$$

which is obviously worse than the optimal rate $\eta^{\frac{\alpha}{\alpha+1}}$. In Faridani and Rieder [28] the authors prove that the algorithm of semi-discrete filtered backprojection applied to the two-dimensional Radon transform is optimal. Since that algorithm is contained in the framework of approximate inverse, as we have seen in section 2.2, see also Rieder, Schuster [102], we have the conjecture that $\widetilde{\mathbf{D}}_{n,d}$ represents a regularization which actually is optimal.

9

Approaches for defect correction

We conclude this part by demonstrating how defect correction methods can be developed, which are adjusted to the method of approximate inverse. By the *defect* we mean the approximation error $\|\mathbf{f} - \mathbf{f}_{\text{app}}\|_{L^2(\Omega,\mathbb{R}^3)}$ of the exact solution $\mathbf{f}$ and the calculated solution $\mathbf{f}_{\text{app}}$. These methods are supposed to improve the accuracy of the reconstruction procedure outlined in the previous chapters. As we pointed out in Remark 6.4 the Doppler transform has a non-trivial null space. When we look at the convergence Theorem 8.3 we moreover see that the reconstruction error depends on how large the parts of the sought solution $\mathbf{f}$ are, which lie in the kernel of $\mathbf{D}_j$. Hence, each approximate solution $\mathbf{f}_{\text{app}}$ of problem (6.10) contains unwished parts from the null spaces of $\mathbf{D}_j$. Since these parts might cause great approximation errors $\|\mathbf{f} - \mathbf{f}_{\text{app}}\|_{L^2(\Omega,\mathbb{R}^3)}$, we are interested in approximating the error $\mathbf{f}_{\text{def}} = \mathbf{f} - \mathbf{f}_{\text{app}}$ and computing a defect correction via

$$\mathbf{f}_{\text{new}} = \mathbf{f}_{\text{app}} + \mathbf{f}_{\text{def}} \,.$$

Over the last few years different approaches for defect correction methods have been developed. References are DEREVTSOV ET AL. [23], SCHUSTER [112] und ANDERSSON [4]. Starting point of the first two articles is the HELMHOLTZ-decomposition

$$\mathbf{f} = \mathbf{f}^{\text{s}} + \nabla v \,, \qquad \nabla \cdot \mathbf{f}^{\text{s}} = 0 \,,$$

where the exact flow is supposed to be incompressible $\nabla \cdot \mathbf{f} = 0$. Direct methods compute projections $\mathbf{f}_{\text{app}}$ of $\mathbf{f}$ onto a subspace spanned by solenoidal fields, see DEREVTSOV, KASHINA [22]. KAZANTSEV AND BUKGHEIM [57] derived the singular value decomposition in two dimensions which also can be used to establish direct methods of defect correction. Indirect methods calculate the potential v as solution of a boundary value problem. Yet, in 1992 JUHLIN [50] showed that the exact field $\mathbf{f}$ can be obtained by solving a boundary value problem, if only the curl and divergence of $\mathbf{f}$ as well as the normal component at the boundary are known. These are the methods to be described in this chapter.

Andersson [4] suggests another approach. He gains more information of $\mathbf{f}$ by calculating higher order moments of the velocity spectrum $\mathrm{d}S$ (6.2). Remember that the first moment of $\mathrm{d}S$ is the Doppler transform. If we had all moments of $\mathrm{d}S$ available, $\mathbf{f}$ would be uniquely determined up to a small ambiguity, see SPARR ET AL. [120].

In section 9.1 we introduce elliptic Dirichlet and Neumann problems which have potentials v as unique solution and show that the gradient ∇v can be used to calculate a better approximation $\mathbf{f}_{\text{new}}$. We further compare the two arising methods relying on the Dirichlet problem on the one hand and on the Neumann problem on the other hand. A concrete method to solve the Neumann problem is suggested in section 9.2, where $\mathbf{f}_{\text{app}}$ is the approximate inverse for the two-dimensional inversion problem. In that case we need the measured data $\Psi_{p,q}\,\mathbf{Df}$ only and not the reconstruction $\mathbf{f}_{\text{app}}$ itself to compute $\mathbf{f}_{\text{def}}$. A brief treatise of the Dirichlet problem is subject of section 9.3. We note that all considerations are done for two-dimensional Doppler tomography, though they apply also for the three-dimensional case.

9.1 Potentials as solutions of elliptic boundary value problems

We assume $\Omega \subset \mathbb{R}^N$, $N = 2, 3$ to be a bounded domain with piecewise smooth boundary $\partial\Omega$, e.g. $\Omega = B_1(0)$ with $\partial\Omega = S^{N-1}$. Further suppose that the flux $\mathbf{f} \in L^2(\Omega, \mathbb{R}^N)$ is incompressible with normal flow $\xi \in L^2(\partial\Omega)$ at the boundary, i.e.

$$\nabla \cdot \mathbf{f} = 0\,, \qquad \nu_x \cdot \mathbf{f}_{|\partial\Omega} = \xi\,, \tag{9.1}$$

where ν_x is the unit outer normal vector at $\partial\Omega$. By means of the Gauss theorem, this implies

$$\int_{\partial\Omega} \xi(x)\,\mathrm{d}s_x = 0\,. \tag{9.2}$$

E.g. $\xi = 0$ physically means that we have a tangential flow at the boundary and that there is no flux through $\partial\Omega$. Furthermore, we assume that an approximate solution $\mathbf{f}_{\text{app}} \in L^2(\Omega, \mathbb{R}^N)$ of (6.10) is given. E.g. $\mathbf{f}_{\text{app}}$ could be the approximate inverse

$$\mathbf{f}_{\text{app}} = \widetilde{\mathbf{D}}_{n,d}\,\Psi_{p,q,r}\,\mathbf{Df}\,. \tag{9.3}$$

Our starting point is the Helmholtz decomposition, which has been deduced by HELMHOLTZ [44], see also WEYL [130]. According to this decomposition, the approximate solution $\mathbf{f}_{\text{app}} \in L^2(\Omega, \mathbb{R}^N)$ can be written as a sum of a solenoidal part $\mathbf{f}^{\text{s}}$ and an irrotational part ∇v

$$\mathbf{f}_{\text{app}} = \mathbf{f}^{\text{s}} + \nabla v\,, \qquad \nabla \cdot \mathbf{f}^{\text{s}} = 0\,. \tag{9.4}$$

The splitting (9.4) becomes unique, if $v = 0$ on $\partial\Omega$ what we presume whenever using (9.4).

The two approaches for defect correction methods outlined in [23] consist of first solving corresponding boundary value problems to obtain a potential v and then subtracting ∇v from $\mathbf{f}_{\text{app}}$, that is $\mathbf{f}_{\text{def}} = -\nabla v$. The first approach calculates the potential v of the decomposition (9.4). To distinguish the two potentials of the methods, we denote v by $v^{(1)}$. Applying the divergence operator to (9.4) and taking into account that v vanishes at the boundary, yields

$$\begin{cases} \Delta v^{(1)} = \nabla \cdot \mathbf{f}_{\text{app}} \text{ in } \Omega\,, \\ \quad v^{(1)} = 0 \text{ on } \partial\Omega\,. \end{cases} \tag{9.5}$$

The unique solution $v^{(1)}$ of (9.5), which coincides with v from (9.4), is used to compute the corrected field

$$\mathbf{f}_{\text{new}}^{(1)} := \mathbf{f}_{\text{app}} - \nabla v^{(1)}\,. \tag{9.6}$$

Since $\nabla v^{(1)} \in \mathsf{N}(\mathbf{D})$, one hopes for a better result $\mathbf{f}_{\text{new}}^{(1)}$.

The second approach relies on the solution of the Neumann problem

$$\begin{cases} \Delta v^{(2)} = \nabla \cdot \mathbf{f}_{\text{app}} \text{ in } \Omega\,, \\ \partial_{\nu_x} v^{(2)} = \nu_x \cdot \mathbf{f}_{\text{app}} - \xi \text{ on } \partial\Omega\,. \end{cases} \tag{9.7}$$

The solution $v^{(2)}$ is unique up to a constant since

$$\int_{\partial\Omega} \Big((\nu_x \cdot \mathbf{f}_{\text{app}})(x) - \xi(x) \Big)\, \mathrm{d}s_x = \int_{\Omega} (\nabla \cdot \mathbf{f}_{\text{app}})(x)\, \mathrm{d}x$$

by the Gauss theorem and (9.2). Thus $\nabla v^{(2)}$ is uniquely determined and the field

$$\mathbf{f}_{\text{new}}^{(2)} := \mathbf{f}_{\text{app}} - \nabla v^{(2)} \tag{9.8}$$

well-defined. Since $\nu_x \cdot \mathbf{f}_{\text{new}}^{(2)}$ equals ξ on $\partial\Omega$ and $\nabla \cdot \mathbf{f}_{\text{new}}^{(2)} = 0$, this method promises an improvement in accuracy too.

First, we state that both methods in fact lead to better reconstruction results.

Lemma 9.1. *If* $\nabla v^{(\ell)} \neq 0$, $\ell = 1, 2$, *then*

$$\|\mathbf{f} - \mathbf{f}_{\text{new}}^{(\ell)}\|^2_{L^2(\Omega,\mathbb{R}^N)} = \|\mathbf{f} - \mathbf{f}_{\text{app}}\|^2_{L^2(\Omega,\mathbb{R}^N)} - \|\nabla v^{(\ell)}\|^2_{L^2(\Omega,\mathbb{R}^N)} < \|\mathbf{f} - \mathbf{f}_{\text{app}}\|^2_{L^2(\Omega,\mathbb{R}^N)}\,.$$

Proof. We outline the proof for $\ell = 2$. The case $\ell = 1$ is handled accordingly and can be found as Lemma 2.2 in [23].
A straightforward calculation shows

$$\begin{aligned} \|\mathbf{f} - \mathbf{f}_{\text{app}}\|^2_{L^2(\Omega,\mathbb{R}^N)} &= \|\mathbf{f} - \mathbf{f}_{\text{new}}^{(2)} - \nabla v^{(2)}\|^2_{L^2(\Omega,\mathbb{R}^N)} \\ &= \|\mathbf{f} - \mathbf{f}_{\text{new}}^{(2)}\|^2_{L^2(\Omega,\mathbb{R}^N)} - 2\,\langle \mathbf{f} - \mathbf{f}_{\text{new}}^{(2)}, \nabla v^{(2)} \rangle_{L^2(\Omega,\mathbb{R}^N)} + \|\nabla v^{(2)}\|^2_{L^2(\Omega,\mathbb{R}^N)}\,. \end{aligned}$$

Taking into account that $\nabla \cdot (\mathbf{f} - \mathbf{f}^{(2)}_{\text{new}}) = 0$ in Ω as well as $\nu_x \cdot (\mathbf{f} - \mathbf{f}^{(2)}_{\text{new}}) = \xi - \xi = 0$ on $\partial\Omega$, an application of the theorem of Gauss-Ostrogradsky yields

$$\langle \mathbf{f} - \mathbf{f}^{(2)}_{\text{new}}, \nabla v^{(2)} \rangle_{L^2(\Omega,\mathbb{R}^N)} = \int_\Omega (\mathbf{f} - \mathbf{f}^{(2)}_{\text{new}})(x) \cdot \nabla v^{(2)}(x) \, \mathrm{d}x$$

$$= \int_{\partial\Omega} \nu_x \cdot (\mathbf{f} - \mathbf{f}^{(2)}_{\text{new}})(x) \, \mathrm{d}s_x - \int_\Omega (\nabla \cdot (\mathbf{f} - \mathbf{f}^{(2)}_{\text{new}}))(x) \, v^{(2)}(x) \, \mathrm{d}x = 0 \, .$$

Thus, we have

$$\|\mathbf{f} - \mathbf{f}_{\text{app}}\|^2_{L^2(\Omega,\mathbb{R}^N)} = \|\mathbf{f} - \mathbf{f}^{(2)}_{\text{new}}\|^2_{L^2(\Omega,\mathbb{R}^N)} + \|\nabla v^{(2)}\|^2_{L^2(\Omega,\mathbb{R}^N)} \, .$$

□

So far, Lemma 9.1 proves that we found two possibilities to enhance the reconstruction accuracy. But what is the differnce between the two methods and which one is to prefer? To analyze the methods, we introduce the errors

$$\mathbf{e}^{(\ell)} = \mathbf{f} - \mathbf{f}^{(\ell)}_{\text{new}} \, , \qquad \ell = 1, 2$$

and denote the difference by $\bar{\mathbf{e}} = \mathbf{e}^{(2)} - \mathbf{e}^{(1)}$. Lemma 9.2 states that $\bar{\mathbf{e}}$ is a harmonic field being L^2-orthogonal to $\mathbf{e}^{(2)}$.

Lemma 9.2. *a) We have that* $\bar{\mathbf{e}} = \nabla \bar{h}$ *for a harmonic function* $\bar{h}$.
b) The fields $\bar{\mathbf{e}}$ and $\mathbf{e}^{(2)}$ are *L^2-orthogonal to each other,*

$$\langle \bar{\mathbf{e}}, \mathbf{e}^{(2)} \rangle_{L^2(\Omega,\mathbb{R}^N)} = 0 \, .$$

c) We have

$$\|\mathbf{e}^{(2)}\|^2_{L^2(\Omega,\mathbb{R}^N)} = \|\mathbf{e}^{(1)}\|^2_{L^2(\Omega,\mathbb{R}^N)} - \|\bar{\mathbf{e}}\|^2_{L^2(\Omega,\mathbb{R}^N)} \leq \|\mathbf{e}^{(1)}\|^2_{L^2(\Omega,\mathbb{R}^N)} \, .$$

Proof. From the settings (9.6), (9.8) and the boundary value problems (9.5), (9.7), it follows that $\nabla \times \bar{\mathbf{e}} = 0 = \nabla \cdot \bar{\mathbf{e}}$ in Ω. The first identity implies $\bar{\mathbf{e}} = \nabla \bar{h}$ for $\bar{h} \in H^1(\Omega)$, the latter one yields $\Delta \bar{h} = 0$.
Using part a) and the Gauss-Ostrogradsky theorem implies

$$\langle \bar{\mathbf{e}}, \mathbf{e}^{(2)} \rangle_{L^2(\Omega,\mathbb{R}^N)} = \int_{\partial\Omega} (\nu_x \cdot \mathbf{e}^{(2)})(x) \, \bar{h}(x) \, \mathrm{d}s_x - \int_\Omega (\nabla \cdot \mathbf{e}^{(2)})(x) \, \bar{h}(x) \, \mathrm{d}x = 0$$

because of (9.7). This gives part b).
Part c) is a simple consequence from part b), since $\mathbf{e}^{(1)} = \mathbf{e}^{(2)} - \bar{\mathbf{e}}$. □

From Lemma 9.2, we read that $\mathbf{f}^{(2)}_{\text{new}}$ represents a better correction than $\mathbf{f}^{(1)}_{\text{new}}$. But note that the boundary values $\xi = \nu_x \cdot \mathbf{f}$ must be known to apply this method. A more detailed analysis of the two approaches is found in Schuster [113].

9.2 The Neumann problem

We show in this section an adept algorithm to solve the Neumann problem (9.7) by means of a boundary element method. We deal with the two-dimensional case ($N = 2$) using a slight modification of the approximate inverse as approximate solution $\mathbf{f}_{\text{app}}$. As a special feature, the algorithm of defect correction uses the measured data $\Psi_{p,q}\,\mathbf{Df}$ only and hence can be performed in parallel to the reconstruction process. For the sake of a better readability, we write $v = v^{(2)}$. We essentially follow the outlines of [112][1].

9.2.1 A boundary element method for the Neumann problem

First, we apply a boundary element method (BEM) to the Neumann problem (9.7) which results in an integral equation of second kind. Denoting with u^* the fundamental solution of the Laplacian

$$\Delta u^*(x,y) = -\delta(x-y)\,, \qquad x,y \in \Omega$$

with Dirac's delta distribution $\delta(x)$, we can identify $v_{|\partial\Omega}$ as solution of an integral equation of second kind

$$\Big(\frac{1}{2}I+\mathbf{K}\Big)v(y) = \int_{\partial\Omega} u^*(x,y)\,(\nu_x\cdot\mathbf{f}_{\text{app}})(x)\,\mathrm{d}s_x - \int_{\Omega} u^*(x,y)\,(\nabla\cdot\mathbf{f}_{\text{app}})(x)\,\mathrm{d}x \quad (9.9)$$

for $y \in \partial\Omega$, see e.g. CHEN, ZHOU [14]. The integral operator $\mathbf{K}$ is the *double layer potential*

$$\mathbf{K}v(y) = \int_{\partial\Omega} \frac{\partial u^*}{\partial \nu_x}(x,y)\,v(x)\,\mathrm{d}s_x\,. \tag{9.10}$$

To eliminate the differentiation of $\mathbf{f}_{\text{app}}$ in (9.9) we apply an integration by parts and get

$$\Big(\frac{1}{2}I+\mathbf{K}\Big)v(y) = \int_{\Omega} \mathbf{f}_{\text{app}}(x)\cdot\nabla_x u^*(x,y)\,\mathrm{d}x =: z(y)\,. \tag{9.11}$$

From (9.11), we get the Dirichlet data $v(y)$, $y \in \partial\Omega$. The solution v of (9.7) is then given by

$$v(y) = z(y) - \int_{\partial\Omega} \frac{\partial u^*}{\partial \nu_x}(x,y)\,v(x)\,\mathrm{d}s_x \tag{9.12}$$

for $y \in \Omega$, where $v_{|\partial\Omega}$ comes from solving (9.11).

The defect correction method reads then as follows:

- Solve the integral equation (9.11) to get the Dirichlet data of v.
- Compute $v(y)$ for $y \in \Omega$ by (9.12).

---

[1] The figures in this section were reproduced from T. SCHUSTER, *Defect correction in vector field tomography: detecting the potential part using BEM and implementation of the method*, Inverse Problems, 21 (2005), pp. 75–91. Copyright ©2005 IOP Publishing Limited. Reprinted with permission.'

- Compute the corrected field

$$\mathbf{f}^{(2)}_{\text{new}} = \mathbf{f}_{\text{app}} - \nabla v\,.$$

Unfortunately, the Newton potential

$$z(y) = \int_{\Omega} \mathbf{f}_{\text{app}}(x) \cdot \nabla_x u^*(x,y)\,\mathrm{d}x \tag{9.13}$$

reduces the efficiency of the BEM, since the evaluation of (9.13) is of higher order than those of integrals over the boundary $\partial\Omega$. Thus, we aim to calculate approximations of (9.13) analytically using the special structure of our reconstruction $\mathbf{f}_{\text{app}}$.

9.2.2 The computation of the Newton potentials

As we only dealt with the three-dimensional case since Chapter 7, we start with a brief re-formulation of the semi-discrete approximate inverse for the 2D Doppler transform $\Psi_{p,q}\mathbf{D}$ with $\Psi_{p,q}$ as in (6.9). We further consider the case $\Omega = \Omega^2$ with $\partial\Omega^2 = S^1$.

The semi-discrete approximate inverse $\widetilde{\mathbf{D}}_{n,d} : \mathbb{R}^n \to L^2(\Omega^2, \mathbb{R}^2)$, $n = 2pq$, is defined via

$$(\widetilde{\mathbf{D}}_{n,d} w)_j(x) = \sum_{\substack{k\in\mathbb{Z}^2\\ \|k\|\le d-1}} \langle w, G_{p,q}\,\Psi_{p,q}\,T_j^{d,k} v^j\rangle_{\mathbb{R}^n}\, B_{d,k}(x)\,.$$

Here,

$$B_{d,k}(x) = B(d\,x - k)\,, \quad d > 0\,, \quad k \in \mathbb{Z}^2$$

with the tensor product splines $B = b \otimes b$ according to (7.13), $T_j^{d,k}$ from (8.2) and the Gramian matrix $G_{p,q} \in \mathbb{R}^{n\times n}$ with respect to the splines (7.1)

$$G_{p,q} = \frac{2\pi}{pq} I_{n,n}\,.$$

For $w = \Psi_{p,q}\,\mathbf{D}\mathbf{f}$, $\mathbf{f} \in H^\alpha(\Omega^2, \mathbb{R}^2)$ and $\alpha > 1/2$, the inner products compute as

$$\begin{aligned}\langle \Psi_{p,q}\,\mathbf{D}\mathbf{f}, G_{p,q}\,\Psi_{p,q}\,T_j^{d,k} v^j\rangle_{\mathbb{R}^n} \\ = \frac{2\pi\,d^2}{pq} \sum_{l=0}^{p-1}\sum_{i=-q}^{q-1} \mathbf{D}\mathbf{f}(\varphi_l, s_i)\, v^j\Big(\varphi_l, d\,s_i - \langle k, \omega(\varphi_l)\rangle\Big)\,.\end{aligned} \tag{9.14}$$

The reconstruction kernel v^j is assumed to be associated with the mollifier $e(x_1,x_2) = \mathrm{p}(x_1,x_2)$ with p as in (7.16) for $\nu = 2$. This gives

$$v^j(\varphi, s) = -\pi^{-2}\,\omega_j^\perp(\varphi)\,\tilde{\phi}(s)\,, \qquad j = 1,2 \tag{9.15}$$

with

$$\tilde{\phi}(s) = \begin{cases} s^2\,(6-4s^2)+\kappa_1 : |s|<1\,, \\ s^{-2}\,{}_2\mathrm{F}_1(1,3/2;4;s^{-2})+\kappa_2 : |s|\geq 1\,, \end{cases}$$

where $\kappa_1 = -(6+8\pi)/15$ and $\kappa_2 = (33-16\pi)/30$. This in fact is representation (7.35) with $\mathrm{q}(a) = 1$ and $\nu = 2$.

In order to get approximations to the Newton potentials $z(y)$ in an explicit form, we slightly simplify the operator $\widetilde{\mathbf{D}}_{n,d}$. Observing that

$$\langle w, G_{p,q}\,\Psi_{p,q}\,T_j^{d,k} v^j\rangle_{\mathbb{R}^n} = (\widetilde{\mathbf{D}}_{n,d} w)_j(d^{-1}\,k)\,, \quad d>0\,, \quad k\in\mathbb{Z}^2$$

we deduce that $\mathbf{B}^j_{n,d} : \mathbb{R}^n \to L^2(\Omega^2)$ with

$$\mathbf{B}^j_{n,d} w(x) = \frac{2\pi\,d^2}{pq}\sum_{l=0}^{p-1}\sum_{i=-q}^{q-1} w_{l,i}\,v^j\Big(\varphi_l, d\,(s_i - \langle x, \omega(\varphi_l)\rangle)\Big)$$

is a reasonable replacement for $(\widetilde{\mathbf{D}}_{n,d} w)_j(x)$. Actually $(\widetilde{\mathbf{D}}_{n,d} w)_j(x)$ emerges from $\mathbf{B}^j_{n,d} w(x)$ using piecewise linear interpolation with respect to the nodes $\{d^{-1}\,k\}$. In contrast to $(\widetilde{\mathbf{D}}_{n,d} w)_j$, the functions $\mathbf{B}^j_{n,d} w$ have a continuous derivative due to the smoothness of v^j. For the remainder of this section, we put

$$(\mathbf{f}_{\mathrm{app}})_j = \mathbf{B}^j_{n,d}\,\Psi_{p,q}\,\mathbf{Df}\,. \tag{9.16}$$

We focus now on the Newton potentials $z(y)$. A simple calculation shows

$$\begin{aligned} z(y) &= \int_{\Omega^2} \mathbf{f}_{\mathrm{app}}(x)\cdot\nabla_x u^*(x,y)\,\mathrm{d}x = \sum_{j=1}^{2}\int_{\Omega^2} (\mathbf{B}^j_{n,d}\,\Psi_{p,q}\,\mathbf{Df})(x)\,\frac{\partial u^*}{\partial x_j}(x,y)\,\mathrm{d}x \\ &= \sum_{j=1}^{2}\langle \mathbf{B}^j_{n,d}\,\Psi_{p,q}\,\mathbf{Df}, \frac{\partial u^*}{\partial x_j}(\cdot,y)\rangle_{L^2(\Omega^2)} \\ &= \sum_{j=1}^{2}\langle \Psi_{p,q}\,\mathbf{Df}, (\mathbf{B}^j_{n,d})^*\Big\{\frac{\partial u^*}{\partial x_j}(\cdot,y)\Big\}\rangle_{\mathbb{R}^n}\,, \end{aligned} \tag{9.17}$$

where the adjoint operator $(\mathbf{B}^j_{n,d})^* : L^2(\Omega^2)\to\mathbb{R}^n$ of $\mathbf{B}^j_{n,d}$ reads

$$[(\mathbf{B}^j_{n,d})^* f]_{l,i} = \frac{2\pi\,d^2}{pq}\int_{\Omega^2} f(x)\,v^j\Big(\varphi_l, d\,(s_i - \langle x, \omega(\varphi_l)\rangle)\Big)\,\mathrm{d}x\,. \tag{9.18}$$

The fundamental solution u^* of the two-dimensional Laplacian is given by

$$u^*(x,y) = -\frac{1}{2\pi}\,\log(\|x-y\|) \tag{9.19}$$

yielding

$$\frac{\partial u^*}{\partial x_j}(x,y) = -\frac{1}{2\pi}\,\frac{x_j - y_j}{\|x-y\|^2}\,.$$

To calculate $z(y)$, we use not only the special form of $\mathbf{f}_{\text{app}}$ (9.16), but also the special structure of the kernel v^j. Obviously the integrals

$$(\mathbf{B}^j_{n,d})^*\Big\{\frac{\partial u^*}{\partial x_j}(\cdot,y)\Big\} = -\frac{d^2}{pq}\int_{\Omega^2}\frac{x_j-y_j}{\|x-y\|^2}\,v^j\Big(\varphi, d\,(s-\langle x,\omega(\varphi)\rangle)\Big)\,\mathrm{d}x\,, \quad (9.20)$$

which appear in (9.17), are singular.

To approximate integrals as in (9.20), we use representation (9.15) and investigate the integrals

$$\Upsilon^j_d f(\varphi,s) := -\frac{d^2}{\pi^2}\,\omega_j^{\perp}(\varphi)\int_{\Omega^2} f(x)\,\tilde{\phi}\Big(d\,(s-\langle x,\omega(\varphi)\rangle)\Big)\,\mathrm{d}x \qquad (9.21)$$

for arbitrary $f \in \mathcal{C}^\infty(\Omega^2)$ with $\varphi \in [0,2\pi]$, $s \in [-1,1]$ and $\tilde{\phi}$ from (9.15). In order to analyze the behavior of $\Upsilon^j_d f$ for large values d, we define the functions

$$F_d(s) = d\,\tilde{\phi}(d\,s)\,, \quad s \in [-2,2]\,, \quad d > 0\,. \qquad (9.22)$$

Figure 9.1 shows a plot of $d^{-1}\cdot F_d$ for different values of d. We notice that $d^{-1}\cdot F_d$ has three extreme values and is almost constant ($=\kappa_2$) for s large. If d is increased, the graph consists almost only of the three extreme values and is constant everywhere else. This observation is important for our searched for approximation of (9.21).

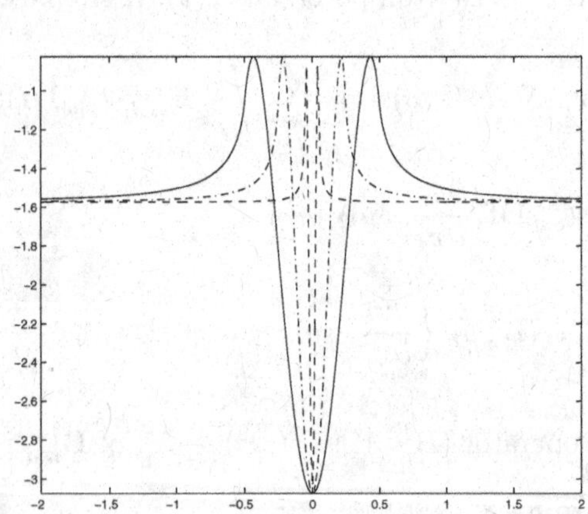

Fig. 9.1. Plot of $d^{-1}\cdot F_d$ for $d=2$ (solid curve), $d=4$ (dashed-dotted curve) and $d=20$ (dashed curve).

At first, we localize the three extreme values of $d^{-1}\cdot F_d$.

Lemma 9.3. *The functions $d^{-1}\cdot F_d$ have two global maximum points at $\mathcal{P}_{1,2}\Big(\pm\frac{\sqrt{3}}{2}\,d^{-1}, (\kappa_1+\frac{9}{4})\Big)$ and a global minimum at $\mathbf{B}(0\,,\,\kappa_1)$.*

Proof. Differentiating F_d gives $F_d'(s) = d^2\,\tilde{\phi}'(d\,s)$ where

$$\tilde{\phi}'(s) = \begin{cases} 12\,s - 16\,s^3\,, & |s| \le 1\,, \\ -\frac{1}{2}\,s^{-3}\,{}_2\mathrm{F}_1(1,3/2;4;s^{-2}) - \frac{1}{2}\,s^{-5}\,{}_2\mathrm{F}_1'(1,3/2;4;s^{-2})\,, & |s| > 1\,. \end{cases} \tag{9.23}$$

From (9.23) we see that $\tilde{\phi}'$ has the zeros $s = 0$ and $s = \pm\sqrt{3}/2$ in $[-1,1]$ and that $\tilde{\phi}$ is monotone decreasing, if $s > 1$, and increasing, if $s < 1$. Thus, there are only critical points in $[-1,1]$. A little bit of analysis shows that we have a minimum point at $s = 0$ and maximum points at $s = \pm\sqrt{3}/2$. □

Lemma 9.3 tells us that the three extreme points of F_d lie in the Interval $\mathcal{I}_d = [-\frac{\sqrt{3}}{2}d^{-1}, \frac{\sqrt{3}}{2}d^{-1}]$ which is concentrated about 0 as $d \to \infty$. Because of $|s - \langle x, \omega(\varphi)\rangle| \le 2$ we are interested in the behavior of F_d in $[-2,2]$ for large values of d. To this end, we take an arbitrary Schwartz function $v \in \mathcal{D}(-2,2)$ with compact support in $[-2,2]$ and compute

$$\int_{-2}^{2} \lambda(s)\,F_d(s)\,ds$$

for large d. Note, that for sufficiently large d we have that $\mathcal{I}_d \subset [-2,2]$. Thus, we can divide $[-2,2]$ into subintervals $\mathcal{I}_d$ and $[-2,2]\backslash\mathcal{I}_d$, apply the trapezoidal sum corresponding to the nodes $s_-^d = -\sqrt{3}d^{-1}/2$, $s = 0$ and $s_+^d = +\sqrt{3}d^{-1}/2$ and obtain

$$\begin{aligned}
\int_{-2}^{2} \lambda(s)\,F_d(s)\,\mathrm{ds} &= \int_{\mathcal{I}_d} \lambda(s)\,F_d(s)\,\mathrm{ds} + \int_{[-2,2]\backslash\mathcal{I}_d} \lambda(s)\,F_d(s)\,\mathrm{ds} \\
&\approx \frac{\sqrt{3}}{2}\,d^{-1}\Big\{\frac{\kappa_1 + 9/4}{d^{-1}}\,\lambda(s_-^d) + \frac{\kappa_1}{d^{-1}}\,\lambda(0) + \frac{\kappa_1 + 9/4}{d^{-1}}\,\lambda(s_+^d)\Big\} \\
&\quad + \int_{[-2,2]\backslash\mathcal{I}_d} \lambda(s)\,F_d(s)\,\mathrm{ds} \\
&\approx \frac{\sqrt{3}}{2}\,d^{-1}\Big\{\frac{\kappa_1 + 9/4}{d^{-1}}\,\lambda(s_-^d) + \frac{\kappa_1}{d^{-1}}\,\lambda(0) + \frac{\kappa_1 + 9/4}{d^{-1}}\,\lambda(s_+^d)\Big\} \\
&\quad + \frac{\kappa_2}{d^{-1}} \int_{-2}^{2} \lambda(s)\,\mathrm{ds}\,.
\end{aligned}$$

Note that for $d \to \infty$ the approximation error in the calculation above gets arbitrarily small. We used also the fact that F_d is almost constant equal to $d\,\kappa_2$ outside $\mathcal{I}_d$ and that $[-2,2]\backslash\mathcal{I}_d \approx [-2,2]$ for large values of d. The considerations above inspire for large $d > 0$ the approximation

$$F_d(s) \approx F_d^0(s) := \frac{\sqrt{3}}{2}\Big(\kappa_1\,\delta(s) + \frac{4\,\kappa_1 + 9}{4}\,(\delta(s - s_-^d) + \delta(s + s_+^d))\Big) + d\,\kappa_2\,, \tag{9.24}$$

where δ denotes Dirac's delta distribution. Thus, we may replace $F_d(s)$ by $F_d^0(s)$ as $d \to \infty$ and get a nice relationship to the Radon transform $\mathbf{R}$.

Lemma 9.4. *Let $f \in C^\infty(\Omega^2)$, $s \in [-1,1]$ and $\varphi \in [0, 2\pi]$. Then*

$$\int_{\Omega^2} f(x)\, F_d^0(s - \langle x, \omega(\varphi)\rangle)\, \mathrm{d}x = \mathbf{R}_d f(\varphi, s)\,, \tag{9.25}$$

where the integral operator $\mathbf{R}_d$ is defined via

$$\mathbf{R}_d f(\varphi,s) = \frac{\sqrt{3}}{2}\Big(\kappa_1\, \mathbf{R} f(\varphi,s) + \frac{4\,\kappa_1 + 9}{4}\,(\mathbf{R} f(\varphi, s - s_-^d) + \mathbf{R} f(\varphi, s + s_+^d))\Big) \\ + d\,\kappa_2\, \Omega^2(f)$$

and $\Omega^2(f) = \int_{\Omega^2} f(x)\, dx$.

Proof. Assertion (9.25) is an immediate consequence of (9.24) and the definition of the Radon transform (2.17). □

With the help of Lemma 9.4 we get an approximation of $\Upsilon_d^j f$ (9.21).

Corollary 9.5. *For large $d \to \infty$ and $f \in C^\infty(\Omega^2)$ we get the approximation*

$$\Upsilon_d^j f(\varphi, s) = -\frac{d}{\pi^2}\, \omega_j^\perp(\varphi)\, \mathbf{R}_d f(\varphi, s)\,. \tag{9.26}$$

Proof. From (9.21), (9.24) and (9.25) we deduce

$$\begin{aligned} \Upsilon_d^j f(\varphi, s) &= -\frac{d}{\pi^2}\, \omega_j^\perp(\varphi) \int_{\Omega^2} f(x)\, F_d(s - \langle x, \omega(\varphi)\rangle)\, \mathrm{d}x \\ &= -\frac{d}{\pi^2}\, \omega_j^\perp(\varphi) \int_{\Omega^2} f(x)\, F_d^0(s - \langle x, \omega(\varphi)\rangle)\, \mathrm{d}x \\ &= -\frac{d}{\pi^2}\, \omega_j^\perp(\varphi)\, \mathbf{R}_d f(\varphi, s) \end{aligned}$$

as $d \to \infty$. □

Remark 9.6. Statement (9.26) in Corollary 9.5 is to be meant in the sense that

$$\lim_{d\to\infty} |\Upsilon_d^j f(\varphi, s) + \frac{d}{\pi^2}\, \omega_j^\perp(\varphi)\, \mathbf{R}_d f(\varphi, s)| = 0 \quad \text{for all } \varphi \in [0, 2\pi]\,, \quad s \in [-1,1]\,.$$

Using the approximation (9.26), we reduce the dimension of the domain of integration. Instead of the two-dimensional domain Ω^2, we only need to integrate over lines. Moreover, if we are able to calculate the Radon transform of the function f explicitly, we get also an explicit representation of $\Upsilon_d^j f$. Since

$$[(\mathbf{B}_{n,d}^j)^* f]_{l,i} = \frac{2\pi}{pq}\, \Upsilon_d^j f(\varphi_l, s_i)\,, \quad 0 \le l \le p-1\,, \quad -q \le i \le q-1\,,$$

we get an expression for the Newton potential $z(y)$ as $d \to \infty$.

Theorem 9.7. *Let* $(\mathbf{f}_{\text{app}})_j = \mathbf{B}^j_{n,d}\,\Psi_{p,q}\,\mathbf{Df}$ *be the approximate inverse* (9.16) *of* $\Psi_{p,q}\,\mathbf{Df}$ *and* $\mathbf{f} \in H^\alpha(\Omega^2, \mathbb{R}^2)$ *for* $\alpha > 1/2$. *Then, the Newton potential* $z(y)$ *has the representation*

$$z(y) = \int_{\Omega^2} \mathbf{f}_{\text{app}}(x) \cdot \nabla_x u^*(x,y)\,\mathrm{d}x$$

$$= -\frac{2\,d}{\pi p q} \sum_{j=1}^{2} \sum_{l=0}^{p-1} \sum_{i=-q}^{q-1} \mathbf{Df}(\varphi_l, s_i)\,\omega_j^\perp(\varphi_l)\,\mathbf{R}_d\{\partial_{x_j} u^*(\cdot,y)\}(\varphi_l, s_i) \tag{9.27}$$

as $d \to \infty$.

Proof. The proof of Theorem 9.7 follows readily from (9.13), (9.17) and (9.26). □

A short calculation shows

$$\Omega^2\{\partial_{x_j} u^*(\cdot,y)\} = \frac{1}{2}\,y_j\,, \tag{9.28}$$

see [23, Lemma 4.2], where the integral is to be understood in the principal value sense. In that very paper an alternative approach is presented to approximate $z(y)$, replacing the integrand in (9.20) by a function which is bounded in Ω^2. Analytic expressions for

$$\mathbf{R}\{\partial_{x_j} u^*(\cdot,y)\}(\varphi, s) \tag{9.29}$$

have been computed in [112, Lemma A.1], when $\|y\| < 1$.

Remark 9.8. The calculation of $z(y)$ with the help of Theorem 9.7 requires the measured data $\Psi_{p,q}\,\mathbf{Df}$ only, not the reconstruction $\mathbf{f}_{\text{app}}$ itself, which occurs in (9.13). Thus, the computation of the defect correction term ∇v can be done without knowledge of $\mathbf{f}_{\text{app}}$. Moreover, both processes, the reconstruction and the defect correction, can be performed in parallel.
Furthermore we mention that the condition $f \in C^\infty(\Omega^2)$ in Lemma 9.4 and Corollary 9.5 is meaningful since $\partial_{x_j} u^*(\cdot,y) \in C^\infty(\Omega^2\backslash\{y\})$. The singularity in (9.27) appears only, if one of the finitely many lines $L(\varphi_k, s_l)$ contains y and is a weak singularity, see (9.28). In case that $y \in L(\varphi_k, s_l)$ the expressions (9.29) still are finite for $\|y\| < 1$, see [112, Lemma A.1].
The (large) factor d in (9.27) comes from the fact that we have a two-dimensional dilation of the mollifier $d^2\,e(d\,(y-x)) \cdot \delta_j$, but only a one-dimensional dilation in $\tilde{\phi}(s)$, see (9.22).

Thus, we may conclude, that we have a possibility to evaluate $z(y)$ for an *arbitrary* $y \in \Omega^2$. But note that (9.27) is only valid in the special case when we use $\mathbf{f}_{\text{app}} = \mathbf{B}^j_{n,d}\,\Psi_{p,q}\,\mathbf{Df}$ to reconstruct $\mathbf{f}$ and the special reconstruction kernel (7.16). But obviously all the calculations and investigations can be done accordingly, if we use another reconstruction kernel. Unfortunately, formula (9.27) is not of filtered backprojection type and thus more time consuming than the reconstruction process. But since we will use the defect correction method only in case of a small number of data, this drawback is justified by the better reconstruction result.

9.2.3 Numerical results

We use the technique of evaluating $z(y)$ outlined in section 9.2.2 to test the algorithm of defect correction. Hereby, we solve the integral equation (9.9), applying a collocation method approximating the boundary $\Gamma = \partial\Omega^2 = S^1$ by a polygon Γ_h and making a piecewise constant ansatz for v on Γ_h, see figure 9.2. Since the collocation points are inner points of Ω^2, we only need a stable evaluation of $z(y)$ for $\|y\| < 1$ which is given by (9.27). The details of this method are drawn from BEBENDORF, RJASANOW [7, 8].

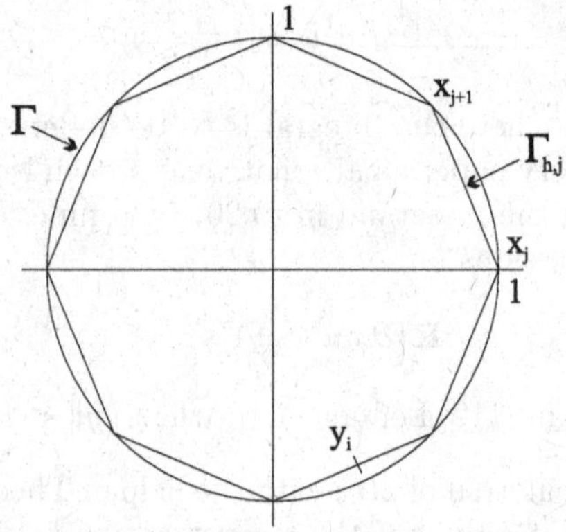

Fig. 9.2. The unit circle $\Gamma = S^1$ with the polygon Γ_h and vertices x_j and x_{j+1}. The collocation points y_j are defined as the center points of an edge $\Gamma_{h,j}$ and hence satisfy $\|y\| < 1$. We make the piecewise constant ansatz $p_{|\Gamma_{h,j}} = c_j$.

Vector fields which fulfill (9.1) on Ω^2 with $\xi = 0$ have the form

$$\mathbf{f} = \nabla \times_\perp w(x_1, x_2) = (-\partial_{x_2} w, \partial_{x_1} w)\,, \qquad w \in H^1(\Omega^2) \tag{9.30}$$

with $-x_1\,\partial_{x_2} w + x_2\,\partial_{x_1} w = 0$ on S^1. Furthermore, we set $\mathbf{f} = 0$ in $\mathbb{R}^2 \setminus \overline{\Omega^2}$. Setting

$$w(x_1, x_2) = -\exp(-x_1^2 - x_2^2)\,, \tag{9.31}$$

we obtain the vector field plotted in figure 7.3. The method of defect correction is of large interest, especially if the amount of data is rather small. Data acquisition might be time consuming, thus only few data can be measured. To adapt this situation, we choose $p = 10$ and $q = 15$. We calculate $\mathbf{f}_{\text{app}}$

applying the method of approximate inverse (9.16) with $d = 4.54$, leading to a relative error

$$\Big(\sum_{j=1}^{2} \|\mathbf{B}_{n,d}^j \Psi_{p,q}\, \mathbf{Df} - \mathbf{f}_j\|_{L^2(\Omega^2)}^2 / \sum_{j=1}^{2} \|\mathbf{f}_j\|_{L^2(\Omega^2)}^2\Big) = 0.165 = 16.5\%\,.$$

A plot of $\mathbf{f}_{\mathrm{app}}$ can be seen in figure 9.3 (left picture). The choice of d might not be optimal, but often it is not possible to determine the optimal value for d, causing further inaccuracies. To compute the defect correction $\mathbf{f}_{\mathrm{new}}^{(2)}$, we first solve the integral equation (9.11) by the mentioned collocation method using 500 collocation points and obtain the Dirichlet data for the searched for potential v. With (9.12) we compute then v and ∇v analytically. Finally, we determine $\mathbf{f}_{\mathrm{new}}^{(2)} = \mathbf{f}_{\mathrm{app}} - \nabla v$ resulting in the relative error

$$\Big(\sum_{j=1}^{2} \|(\mathbf{f}_{\mathrm{new}}^{(2)})_j - \mathbf{f}_j\|_{L^2(\Omega^2)}^2 / \sum_{j=1}^{2} \|\mathbf{f}_j\|_{L^2(\Omega^2)}^2\Big) = 0.1292 = 12.92\%\,,$$

which means an improvement by about 4%. The potential field ∇v is displayed in figure 9.3 (right picture).

Figure 9.4 contains the approximate inverse $\mathbf{f}_{\mathrm{app}}$ of the vector field $\mathbf{f}$ corresponding to

$$w(x_1, x_2) = (1 - \|x\|^2)^2 \cos(x_1 + x_2) \tag{9.32}$$

in (9.30) as well as the potential field ∇v, where we used noise contaminated data with a noise level of 7%. Here, the original error

$$\Big(\sum_{j=1}^{2} \|\mathbf{B}_{n,d}^j \Psi_{p,q}\, \mathbf{Df} - \mathbf{f}_j\|_{L^2(\Omega^2)}^2 / \sum_{j=1}^{2} \|\mathbf{f}_j\|_{L^2(\Omega^2)}^2\Big) = 0.1 = 10\%$$

could be improved to

$$\Big(\sum_{j=1}^{2} \|(\mathbf{f}_{\mathrm{new}}^{(2)})_j - \mathbf{f}_j\|_{L^2(\Omega^2)}^2 / \sum_{j=1}^{2} \|\mathbf{f}_j\|_{L^2(\Omega^2)}^2\Big) = 0.0833 = 8.33\%\,.$$

Note that the improvement is only small, since fields $\nabla \times_\perp w$ are solenoidal and hence we have exact convergence for $d, n \to \infty$ as pointed out in Corollary 8.5.

9.3 The Dirichlet problem

There are several ways to solve the Dirichlet problem (9.5). Certainly, we may apply a BEM also in that case after a reformulation of (9.5) as an integral equation. The Neumann data of $v = v^{(1)}$ are obtained by solving

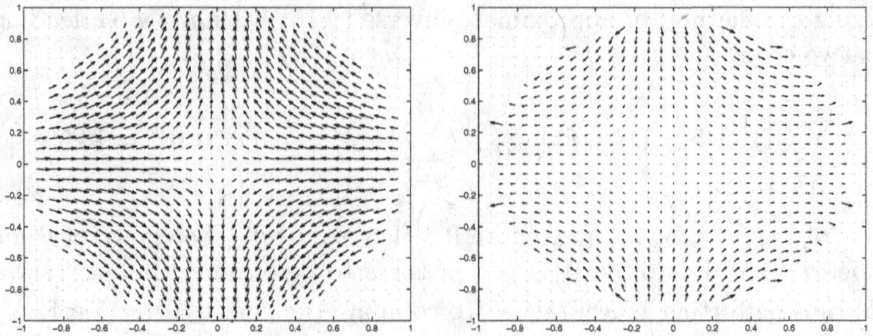

Fig. 9.3. The approximate inverse $\mathbf{f}_{\text{app}}$ of $\mathbf{f}(x_1, x_2) = \nabla\times_\perp w(x_1, x_2) = 2\exp(-x_1^2 - x_2^2)(-x_2, x_1)$ with w from (9.31) (left picture) and the corresponding potential part ∇v (right picture). The structure of $\mathbf{f}$ is illuminated well by the picture to the left. But the plot of the potential part ∇v also shows reconstruction errors near the boundary.

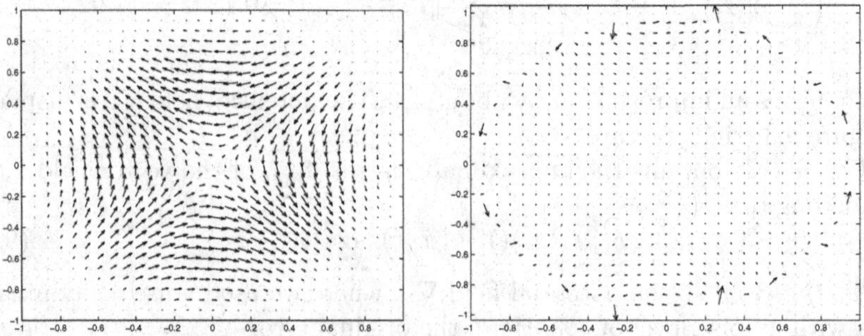

Fig. 9.4. The approximate inverse $\mathbf{f}_{\text{app}}$ of $\mathbf{f} = \nabla\times_\perp w$ with w as in (9.32) (left picture) and the corresponding potential part ∇v (right picture). Again the potential part reveals boundary errors.

$$\Big(\frac{1}{2}I + \widetilde{\mathbf{K}}\Big)v(y) = -\int_\Omega u^*(x,y)\,\Delta p(x)\,\mathrm{d}x \tag{9.33}$$

$$= \int_\Omega \mathbf{f}_{\text{app}}(x)\cdot\nabla_x u^*(x,y)\,\mathrm{d}x - \int_{\partial\Omega} u^*(x,y)\,(\nu_x\cdot\mathbf{f}_{\text{app}})(x)\,\mathrm{d}s_x =: \tilde{z}(y)\,,$$

where $\widetilde{\mathbf{K}}$ is the *single layer potential*

$$\widetilde{\mathbf{K}}v(y) = \int_{\partial\Omega} u^*(x,y)\,\frac{\partial p}{\partial\nu_x}(x)\,\mathrm{d}s_x\,.$$

The evaluation of $\tilde{z}(y)$ can be done as follows. The first integral in (9.33)

$$\int_\Omega \mathbf{f}_{\text{app}}(x)\cdot\nabla_x u^*(x,y)\,\mathrm{d}x\,,$$

is exactly the function $z(y)$, which we considered in section 9.2.2 and is evaluated as indicated in Theorem 9.7. Assuming again that $\Omega = \Omega^2$, the second integral may be treated in the following way. We write the boundary integral in (9.33) first in spherical coordinates

$$\int_{S^1} u^*(x,y)\,(\nu_x \cdot \mathbf{f}_{\mathrm{app}})(x)\,\mathrm{d}s_x = \int_0^{2\pi} u^*(\omega(\vartheta),y)\,\Big(\omega(\vartheta)\cdot \mathbf{f}_{\mathrm{app}}(\omega(\vartheta))\Big)\,\mathrm{d}\vartheta \tag{9.34}$$

$$= \frac{d^2}{\pi^2 pq}\sum_{j=1}^{2}\sum_{l=0}^{p-1}\sum_{i=-q}^{q-1} \mathbf{Df}(\varphi_l, s_i)\,\omega_j^{\perp}(\varphi_l)\times$$

$$\times \int_0^{2\pi} \ln\|\omega(\vartheta)-y\|\,\omega_j(\vartheta)\,\tilde{\phi}\Big(d\,(s_i - \langle\theta(\varphi_l),\omega(\vartheta)\rangle)\Big)\,\mathrm{d}\vartheta\,,$$

where $\omega(\vartheta) = (\cos\vartheta, \sin\vartheta) \in S^1$. Note, that $\nu_x = x$, $x \in S^1$. Since we evaluate (9.34) for values $y \in \Omega^2$ only, there is no singularity in (9.34). Thus, we may apply the trapezoidal rule to get an appropriate approximation. Let

$$\vartheta_\mu = \mu\,\frac{2\pi}{M}, \qquad \mu = 0,\ldots,M\,,$$

be an equispaced discretization in ϑ. Then

$$\int_0^{2\pi} \ln\|\omega(\vartheta)-y\|\,\omega_j(\vartheta)\,\tilde{\phi}\Big(d\,(s_i - \langle\theta(\varphi_l),\omega(\vartheta)\rangle)\Big)\,d\vartheta$$

$$\approx \frac{2\pi}{M}\sum_{\mu=0}^{M} \ln\|\omega(\vartheta_\mu)-y\|\,\omega_j(\vartheta_\mu)\,\tilde{\phi}\Big(d\,(s_i - \langle\theta(\varphi_l),\omega(\vartheta_\mu)\rangle)\Big)\,.$$

To solve (9.33) we can use a collocation method again.

Another approach is described in [23, section 3]. There, the authors suggest a method which uses a conformal mapping $\mathcal{P} : \Omega^2 \to [-2,2]^2$ to transform the Dirichlet problem (9.5) to an equivalent boundary value problem on the square $[-2,2]^2$ which is then solved by means of a finite difference scheme. The drawback of this method is that the mapping $\mathcal{P}$ changes the metric in $[-2,2]^2$ to a non-Euclidean one.

10

Conclusion and perspectives

We transfered the abstract framework done in Part I to the problem of Doppler tomography. The result is a stable inversion scheme of filtered backprojection type which emphasizes all amenities being characteristic for the method of approximate inverse. These are:

- The reconstruction kernels can be pre-computed before the measurement process starts.
- Invariances of the underlying operator help to improve the efficiency of the method. Here we used dilation and translation invariances of the adjoint of the Doppler transform.
- These invariances are further the reason that we have to compute the reconstruction kernel only once. Thus the method is well suited, if one needs to repeat the reconstruction process very often.

The measurement geometry described in Chapter 6 is very time consuming, since the body has to be scanned slice by slice and in three different directions. Hence, from a practical point of view, cone beam data would be more convenient. The application of the method of approximate inverse to the cone beam transform

$$\mathbf{Df}(a,\omega) = \int_0^\infty \langle \omega, \mathbf{f}(a+t\omega)\rangle \, dt\,, \quad \omega \in S^2\,,$$

where $a \in \Gamma \subset (\mathbb{R}^3 \backslash \overline{\Omega})$ is a source point on a given scanning curve, is subject of current research as well as the general development of inversion schemes for the cone beam geometry.

In chapter 9, we emphasized that defect correction methods are necessary to improve the reconstruction accuracy and outlined a method in 2D which relies on the approximate inverse as reconstruction scheme and needs the measured data only to compute the defect correction term. The aim is to extend the presented method to the three-dimensional case, since in that situation we do not have exact convergence for $d, n \to \infty$ in contrast to the two-dimensional case, compare (8.7), (8.9). To solve the according boundary value problems in 3D, fast boundary element methods such as the adaptive cross approximation (BEBENDORF, RJASANOW [8]) could be useful.

Part III

Application to the spherical mean operator

We apply the concept of the distributional approximate inverse, outlined in Chapter 4, to the problem of recovering a function from its spherical means. The elucidations in this part mainly are subject of the article [115][1] by SCHUSTER AND QUINTO. We focus at the inversion of the *spherical mean operator*

$$\mathbf{M}f(a_0, r) = \int_{S(a_0,r)} f(x)\, \mathrm{d}S_n^r(x)\,, \quad S(a_0, r) = \{x \in \mathbb{R}^n : \|x - a_0\| = r\}\,,$$

$a_0 \in A \subset \mathbb{R}^n$. This problem arises in a variety of applications such as thermoacoustic detection of tumours, see KRUGER ET AL. [60], XU, WANG [133], seismics, see ROMANOV [104], and SONAR (SOund in NAvigation and Radiation), see LAVRENTIEV ET AL. [62], LOUIS, QUINTO [74]. A detailed outline of the relation between spherical means and the detection of reflectivity of the earth's surface by SAR (Synthetic Aperture Radar) can be found in CHENEY [15].

This is reason enough to develop inversion schemes for $\mathbf{M}$, where the different models differ from each other by the *center sets* A. NORTON [88] gives an inversion formula for the case where the center set is a circle in a plane, whereas FINCH ET AL. [31] consider the situation where A is the boundary of a bounded, connected and open set in $\mathbb{R}^n$. In RAMM [96] a proof of injectivity of $\mathbf{M}$ for the latter case can be found. DENISJUK [19] considers the general case, where the centers are located on hyperplanes in $\mathbb{R}^n$. He gives an inversion method, which relies on a transform mapping spheres to planes and algebraic reconstruction techniques (ART). The problem of limited data is also treated in this article. ANDERSSON [5] investigates the properties of $\mathbf{M}$ as a bounded mapping between distribution spaces, where the center set is the hyperplane $x_{n+1} = 0$ in $\mathbb{R}^{n+1}$. He deduces an inversion formula with an analogue structure as that of the Radon transform. The computation of reconstruction kernels outlined in Chapter 13 relies on this very formula.

A microlocal analysis for $\mathbf{M}$ has been developed in LOUIS, QUINTO [74] to clarify which singularities of the object to be recovered can be visualized from the given data and which cannot. They prove that only those singularities of f being conormal to the sphere $S(a_0, r)$ can be detected.

This part is organized as follows. The first chapter is dedicated to the investigation of the spherical mean operator $\mathbf{M}$. We shortly summarize the mathematical models in SONAR and SAR (Section 11.1) and the mathematical properties of $\mathbf{M}$ which are proved in ANDERSSON [5] (Section 11.2). Just as in [5] we only consider the case where the center set A is a special hyperplane.

---

[1] 'Some pictures in Part III are taken from T. SCHUSTER AND E.T. QUINTO, *On a regularization scheme for linear operators in distribution spaces with an application to the spherical Radon transform*, SIAM J. Appl. Math., 65 (2005), pp. 1369–1387. Copyright ©2005 Society for Industrial and Applied Mathematics. Reprinted with permission.'

Chapter 11 concludes with the formulation of the method of approximate inverse and its semi-discrete version which is important from a practical point of view. The following chapter deals with the design of an operator adjusted mollifier. The computation of corresponding reconstruction kernels follows in Chapter 13. Chapter 14 shows how the method works with the help of numerical results.

11

The spherical mean operator

After a short treatise about SONAR and SAR, we summarize in Section 11.2 the essential mathematical properties of the spherical mean operator $\mathbf{M}$. Though there are some similarities to the Radon transform $\mathbf{R}$ with respect to its definition and inversion formula, we point out crucial differences when the center set is given by $\{x_{n+1} = 0\} \subset \mathbb{R}^{n+1}$, which causes difficulties in the numerical treatment of that mapping. The most important difference between the two transforms is the fact that $\mathbf{M}$ can neither be formulated as continuous mapping between L^2- nor between Sobolev spaces. Moreover, $\mathbf{M}$ is meaningfully defined on certain spaces of tempered distributions only. Hence to solve the inverse problem

$$\mathbf{M}f = g\,,$$

we need the concepts presented in Chapter 4.

11.1 Spherical means in SONAR and SAR

To detect and visualize objects in the water, one emits ultrasound signals from an antenna and measures reflections. In shallow water the assumption of a constant speed of sound $c(x) = c_0$ is reasonable. A signal $U = U(t, x)$ being emitted from a source a_0 in a domain $A \subset \mathbb{R}^3$ at time $t = 0$ hence generates a spherical wave front. The reflected signal which is received at time t in a_0 thus contains information of all reflections located at a sphere with radius $(t/2)\,c_0$ and center a_0, see Figure 11.1. In Figure 11.1 the center set consists of the line $\{x_2 = 0\}$.

The measured signal is

$$\mathbf{M}f(a_0, r) = \int_{S(a_0,r)} f(x)\,\mathrm{d}S_n^r(x)\,, \qquad r = (t/2)\,c_0\,, \tag{11.1}$$

where $S(a_0, r) := \{x \in \mathbb{R}^3 : \|x - a_0\| = r\}$ and $f(x)$ is the reflectivity. Here $\mathrm{d}S_n^r$ denotes the $n-1$-dimensional surface measure on $S(a_0, r)$. The reflectivity

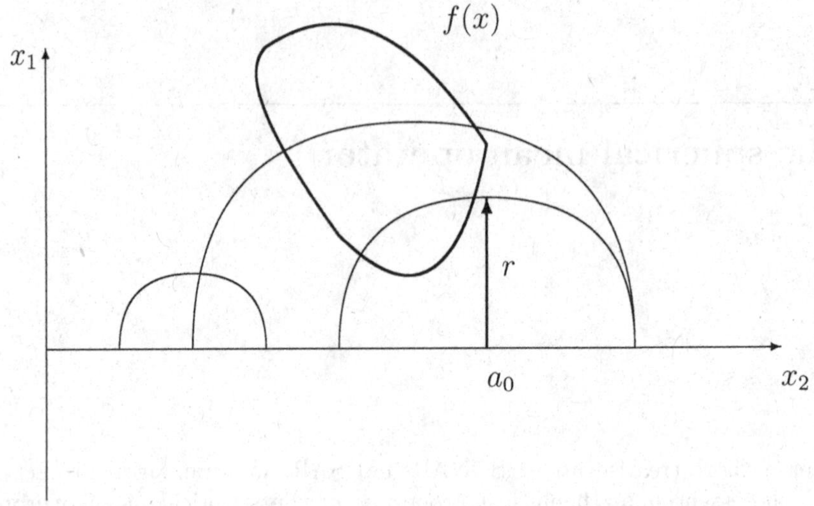

Fig. 11.1. Measurement geometry of SONAR in two dimensions. The x_1- and x_2-axes are switched to be consistent with the definitions in Section 11.2.

essentially depends on the speed of sound $c(x)$, which gives information about objects in the water. The problem to recover objects in water from ultrasound measurements is called SONAR (SOund NAvigation and Radiation). In SONAR the centers a_0 usually are located at the hyperplane $\{x_3 = 0\}$. As can be read in LOUIS, QUINTO [74] the signal $U(t,x)$ propagates according to the acoustic wave equation

$$k^2(x)\,U_{tt} = \Delta U + \delta(t)\,\delta(x - a_0)\,, \qquad a_0 \in A\,. \tag{11.2}$$

Provided that there is no multiple scattering (*Born approximation*) which means a linearization, then the determination of k^2 from the back-scattered signal is equivalent to the reconstruction of k^2 from $\mathbf{M}(k^2)(a_0, r)$, see LAVRIENTIEV ET AL. [62] and ROMANOV [104]. Here, the refraction index k^2 corresponds to the reflectivity f in (11.1).

We have a similar situation in SAR. Here, the aim is to determine objects at the earth's surface by ultrasound signals, where the antenna usually is attached to the wing of an aircraft. To be exact, we would have to consider Maxwell's equations as mathematical model. But for the sake of simplicity, one investigates equation (11.2) or the similar equation

$$U_{tt} = \Delta U + q(x)\,U + \delta(t)\,\delta(x - a_0)\,, \qquad a_0 \in A\,,$$

where $q(x)$ denotes the scatterer. HELLSTEN, ANDERSSON [45] show, how the measured data in SAR can be interpretated as spherical means of the ground reflectivity. CHENEY [15] explains how the signals which are measured at the antenna can be expressed by spherical means of $c^{-1}(x) - c_0^{-1}$ when we take equation (11.2) with $k^2(x) = -1/c^2(x)$ as a starting point.

11.2 Properties of the spherical mean operator

Let $\mathcal{S}(\mathbb{R}^n)$ be the space of rapidly decreasing functions in $\mathbb{R}^n$, i.e. the space of all functions $f \in \mathcal{C}^\infty(\mathbb{R}^n)$ for which the seminorms

$$\mathrm{p}_m(v) = \sup_{|\alpha| \leq m} \sup_{x \in \mathbb{R}^n} (1 + \|x\|^2)^m \, |D^\alpha v(x)| < \infty$$

are finite for all $m \in \mathbb{N}_0$. Here, $\alpha \in \mathbb{N}_0^n$ is a multiindex and $D^\alpha := \partial_{x_1}^{\alpha_1} \cdot \ldots \cdot \partial_{x_n}^{\alpha_n}$ is the differential operator of order $|\alpha| = \alpha_1 + \ldots + \alpha_n$. The system $\{\mathrm{p}_m\}$ induces a local convex topology which turns $\mathcal{S}(\mathbb{R}^n)$ to a *Fréchet space.* Its dual $\mathcal{S}'(\mathbb{R}^n)$ consists of all functionals which are linear and bounded on $\mathcal{S}(\mathbb{R}^n)$. That means that to each $\lambda \in \mathcal{S}'(\mathbb{R}^n)$ there exists a $m \in \mathbb{N}_0$ and a constant $C_m > 0$ satisfying

$$|\langle \lambda, v \rangle_{\mathcal{S}'(\mathbb{R}^n) \times \mathcal{S}(\mathbb{R}^n)}| \leq C_m \, \mathrm{p}_m(v) \qquad \text{for all } v \in \mathcal{S}(\mathbb{R}^n)\,.$$

Thus, each $\lambda \in \mathcal{S}'(\mathbb{R}^n)$ is of finite order. The space $\mathcal{S}'(\mathbb{R}^n)$ is called the space of *tempered distributions.* The following theorem, which can be found e.g. in CONSTANTINESCU [17, Theorem 7.4], characterizes tempered distributions as (weak) derivatives of slowly increasing functions.

Theorem 11.1. *To each $\lambda \in \mathcal{S}'(\mathbb{R}^n)$ there exists a multiindex $\alpha \in \mathbb{N}_0^n$ and a continuous function P_λ of at most polynomial growth, such that*

$$\langle \lambda, v \rangle_{\mathcal{S}'(\mathbb{R}^n) \times \mathcal{S}(\mathbb{R}^n)} = (-1)^{|\alpha|} \int_{\mathbb{R}^n} P_\lambda(x) \, D^\alpha v(x) \, \mathrm{d}x \tag{11.3}$$

for all $v \in \mathcal{S}(\mathbb{R}^n)$.

Obviously, $\mathcal{S}(\mathbb{R}^n) \subset \mathcal{S}'(\mathbb{R}^n)$ and the embedding is dense. As a consequence, the Fourier transform $\mathbf{F}$ can be extended continuously to an isomorphism on $\mathcal{S}'(\mathbb{R}^n)$.

Following the lines in [5] we investigate the particular case where the centers a_0 in (11.1) are located on the hyperplane $\{z \in \mathbb{R}^{n+1} : z_{n+1} = 0\}$. To adapt this very situation we re-define $\mathbf{M}$. The spherical mean operator $\mathbf{M}$ now particularly assigns a function $f \in \mathcal{S}(\mathbb{R}^{n+1})$ to its mean values over all spheres with radius $r \geq 0$ centered about $(z, 0)^\top \in \mathbb{R}^{n+1}$, $z \in \mathbb{R}^n$,

$$\mathbf{M}f(z, r) = \frac{1}{|S^n|} \int_{S^n} f(z + r\,\xi, r\,\eta) \, \mathrm{d}S_n(\xi, \eta) = g(z, r)\,. \tag{11.4}$$

By $|S^n|$ we denote the surface area of the n-dimensional unit sphere $S^n = \{(\xi, \eta)^\top \in \mathbb{R}^{n+1} : \xi \in \mathbb{R}^n,\ \eta \in \mathbb{R},\ \|\xi\|^2 + \eta^2 = 1\} \in \mathbb{R}^{n+1}$, $\mathrm{d}S_n$ is the surface measure on S^n. In contrast to the Radon transform $\mathbf{R}$, the spherical mean operator integrates over n-dimensional hyperspheres S^n instead of n-dimensional planes. We often will use the notation $x = (x', x_{n+1})^\top$ for $x \in$

$\mathbb{R}^{n+1}$, where $x' = (x_1, \ldots, x_n)^\top$ contains the first n components of x and x_{n+1} is the $(n+1)$-st component.

Obviously, $\mathbf{M}$ is not injective, since $\mathbf{M}f = 0$ for each $f \in \mathcal{S}(\mathbb{R}^{n+1})$ being odd with respect to the last variable. COURANT and HILBERT [18] proved that the null space of $\mathbf{M}$ consists of all those functions which are odd in x_{n+1}. Thus, it is reasonable to restrict the domain of $\mathbf{M}$ to the subspace $\mathcal{S}_\mathrm{e}(\mathbb{R}^{n+1}) \subset \mathcal{S}(\mathbb{R}^{n+1})$ of all rapidly decreasing functions being even in x_{n+1},

$$\mathcal{S}_\mathrm{e}(\mathbb{R}^{n+1}) := \{f \in \mathcal{S}(\mathbb{R}^{n+1}) : f(x', -x_{n+1}) = f(x', x_{n+1})\}\,.$$

Unfortunately, $f \in \mathcal{S}_\mathrm{e}(\mathbb{R}^{n+1})$ does not imply that $\mathbf{M}f$ is again a rapidly decreasing function. Even worse: the image $\mathbf{M}f$ in general is neither in $L^2(\mathbb{R}^{n+1})$, nor in $L^1(\mathbb{R}^{n+1})$. This fact is emphasized in Example 11.2 in two dimesnions $(n = 1)$. The image of the characteristic function of two circles under $\mathbf{M}$, that means of a function with compact support, is not even integrable.

Example 11.2. Let $n = 1$ and $\chi_C \in L^2(\mathbb{R}^2)$ be the characteristic function of two disks which are reflected about the x_2-axis

$$\chi_C(x) = \chi_C(x_1, x_2) = \begin{cases} 2\,, & \text{if } \|x - (4,4)\| \le 1 \text{ or } \|x - (4,-4)\| \le 1\,, \\ 0\,, & \text{else}\,. \end{cases} \tag{11.5}$$

Note that χ_C is even with respect to x_2. The picture to the left in Figure 11.2 shows a plot of χ_C for $x_2 > 0$. After some geometric considerations we compute for $z \in \mathbb{R}$ and $r > 0$

$$\mathbf{M}\chi_C(z, r) = \begin{cases} 8\,\pi^{-1}\, r \,\arccos\left(\frac{r^2 + d^2 - 1}{2\,r\,d}\right), & d - 1 \le r \le d + 1\,, \\ 0\,, & \text{else}\,, \end{cases}$$

where $d = \|(z, 0) - (4, 4)\|$. We integrate over spheres with radius $r > 0$; the center set is the line $\{(z, 0) : z \in \mathbb{R}\}$. The picture to the right in Figure 11.2 displays $\mathbf{M}\chi_C$ for (z, r) in $[-35, 35] \times [0, 50]$. Obviously the support of $\mathbf{M}\chi_C$ is not bounded in $\mathbb{R}^2$.

This is a crucial difference compared to the Radon transform. The question arises on which spaces $\mathbf{M}$ can be defined meaningfully as a bounded operator. To answer this question, we first introduce a subspace of $\mathcal{S}(\mathbb{R}^{2n+1})$. Let

$$\mathcal{S}_\mathrm{r}(\mathbb{R}^n \times \mathbb{R}^{n+1}) := \{f \in \mathcal{S}(\mathbb{R}^{2n+1}) : f(z, w) = \check{f}(z, \|w\|) \text{ for } \check{f} \in \mathcal{S}_\mathrm{e}(\mathbb{R}^{n+1})\}.$$

The space $\mathcal{S}_\mathrm{r}(\mathbb{R}^n \times \mathbb{R}^{n+1})$ contains all functions of $\mathcal{S}(\mathbb{R}^{2n+1})$ being radially symmetric in the last $n+1$ variables. Thus we will always understand functions from $\mathcal{S}_\mathrm{r}(\mathbb{R}^n \times \mathbb{R}^{n+1})$ as functions on $\mathbb{R}^n \times \mathbb{R}$ in virtue of the setting $f(z, r) = f(z, w)$, $r = \|w\|$. The reason to use $\mathbb{R}^{n+1}$ for the radial variable rather than $\mathbb{R}$ is that we may apply the Fourier transform to a function from $\mathcal{S}_\mathrm{r}(\mathbb{R}^n \times \mathbb{R}^{n+1})$. This is important in view of Theorem 11.3. Of course the Fourier transform again is radial in the last $n+1$ variables. In analogy to $(\mathcal{S}(\mathbb{R}^n), \mathcal{S}'(\mathbb{R}^n))$ we may

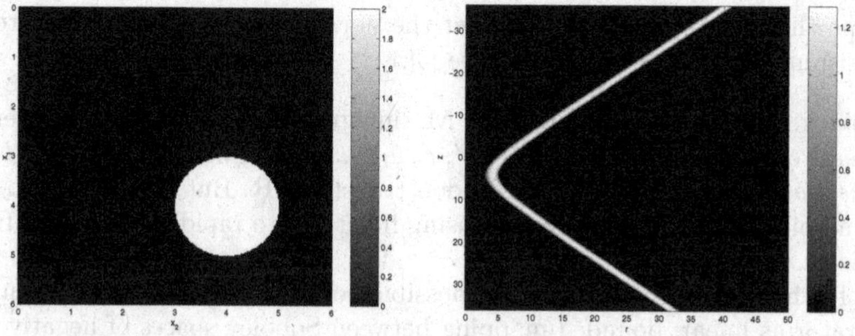

Fig. 11.2. Left picture: Plot of the object function χ_C consisting of two disks reflected about the x_2-axis. Only the part for $x_2 > 0$ is shown. Right picture: Plot of $\mathbf{M}\chi_C$ in $[-35, 35] \times [0, 50]$.

also consider the dual pairings $(\mathcal{S}_e(\mathbb{R}^{n+1}), \mathcal{S}_e'(\mathbb{R}^{n+1}))$, $(\mathcal{S}_r(\mathbb{R}^n \times \mathbb{R}^{n+1}), \mathcal{S}_r'(\mathbb{R}^n \times \mathbb{R}^{n+1}))$. For the sake of a better readability, we set $\mathcal{S}_e := \mathcal{S}_e(\mathbb{R}^{n+1})$ and $\mathcal{S}_r := \mathcal{S}_r(\mathbb{R}^n \times \mathbb{R}^{n+1})$, the notations $\mathcal{S}_e'$, $\mathcal{S}_r'$ are respectively. As a consequence of the considerations made before, we cannot expect that $\mathbf{M}f \in \mathcal{S}_r$, when $f \in \mathcal{S}_e$. But it is easy to show that $\mathbf{M}f \in \mathcal{S}_r'$ for all $f \in \mathcal{S}_e$. Since $\mathcal{S}_e \hookrightarrow \mathcal{S}_e'$ is dense, we even have $\mathbf{M}f \in \mathcal{S}_r'$ whenever $f \in \mathcal{S}_e'$. Further properties are summarized in the following theorem whose proof can be found in ANDERSSON [5] and KLEIN [59].

Theorem 11.3. *The spherical mean operator* $\mathbf{M} : \mathcal{S}_e' \to \mathcal{S}_r'$ *is a linear, continuous mapping which is one-to-one. The range* $\mathsf{R}(\mathbf{M})$ *can be characterized by*

$$\mathsf{R}(\mathbf{M}) = \mathcal{S}_{r,cone}' := \left\{ g \in \mathcal{S}_r' : supp\ \hat{g} \subset \{(\sigma, \varrho) \in \mathbb{R}^n \times [0, \infty) : \varrho \geq \|\sigma\|\} \right\} \subset \mathcal{S}_r'. \tag{11.6}$$

If the Fourier transform of $f \in \mathcal{S}_e'$ *is equal to an integrable function* $\hat{f}(\sigma, \varrho)$, *then the inversion formula*

$$\hat{f}(\sigma, \varrho) = c_n\, |\varrho|\, (\|\sigma\|^2 + \varrho^2)^{(n-1)/2}\, \hat{g}(\sigma, \sqrt{\|\sigma\|^2 + \varrho^2}) \tag{11.7}$$

holds true with $c_n = |S^n|/(2\,(2\,\pi)^n)$ *and* $g = \mathbf{M}f$.
The adjoint operator $\mathbf{M}^* : \mathcal{S}_r \to \mathcal{S}_e$ *has dense range and is given by*

$$\mathbf{M}^* g(x', x_{n+1}) = \int_{\mathbb{R}^n} g\Big(z, \sqrt{\|z - x'\|^2 + x_{n+1}^2}\Big)\, \mathrm{d}z\,. \tag{11.8}$$

We further have

$$\mathbf{F}\,\mathbf{M}^* g(\sigma, \varrho) = \hat{g}(\sigma, \sqrt{\|\sigma\|^2 + \varrho^2})\,. \tag{11.9}$$

Note that the function $\hat{g}$ on the right-hand side of equation (11.7) is a Fourier transform of a function on $\mathbb{R}^{2n+1}$, which is radially symmetric with respect to the last $n + 1$ variables, and so is $\hat{g}$. In the entire Part III, we define the Fourier transform *without* the normalizing factor $(2\pi)^{-n/2}$ to be consistent with ANDERSSON's article [5].

Remark 11.4. The adjoint operator $\mathbf{M}^*$ integrates a function $g \in \mathcal{S}_\mathrm{r}$ over all spheres containing the point $x = (x', x_{n+1})^\top$. That is why $\mathbf{M}^*$ is called *back-projection* just as in case of the Radon transform $\mathbf{R}$. But in contrast to $\mathbf{R}^*$, the adjoint $\mathbf{M}^*$ maps rapidly decreasing functions to rapidly decreasing functions.
We further remark that it is not possible to formulate the spherical mean operator as linear, bounded mapping between Sobolev spaces of negative order. This becomes clear by Figure 11.2 showing that a function, which is not continuous, has a range of large local smoothness or by the estimate

$$\|f\|_{H^\alpha(\mathbb{R}^{n+1})} \leq \sqrt{\frac{|S^n|}{2}}\, \|\mathbf{M}f\|_{H^{\alpha+1/2}(\mathbb{R}^{2n+1})} \tag{11.10}$$

and its proof which can be read in [5, Theorem 3.1]. Note that the Sobolev norm at the right-hand side of the estimate does not need to be finite.
Defining the operator $\mathbf{K} : H^\alpha(\mathbb{R}^{n+1}) \cap \mathcal{S}'_{r,cone} \to H^{\alpha+n}(\mathbb{R}^{2n+1}) \cap \mathcal{S}'_\mathrm{r}$ via $\mathbf{F}\,\mathbf{K}g(\sigma,\varrho) = \sqrt{\varrho^2 - \|\sigma\|^2}\,\varrho^{n-1}\,\hat{g}(\sigma,\varrho)$, the inversion formula (11.7) has the representation

$$f = c_n\, \mathbf{M}^*\, \mathbf{K}\, \mathbf{M} f$$

and hence a structure which is according to the inversion formula (2.19) of $\mathbf{R}$.

We aim to transfer the concepts of Chapter 4 to the problem of solving

$$\mathbf{M}f = g\,. \tag{11.11}$$

To compute reconstruction kernels, we need a solution of

$$\mathbf{M}^* v_\gamma(y) = e_\gamma(y)\,, \qquad e_\gamma(y) \in \mathcal{S}_\mathrm{e}\,. \tag{11.12}$$

Theorem 11.3 tells us that we have the situation described in part b) of Remark 4.3: if $e_\gamma(y) \in \mathsf{R}(\mathbf{M}^*)$, then $v_\gamma(y)$ lies in $\mathcal{S}_\mathrm{r}$. Against this background, the extension lemma [5, Lemma 2.4] is of great importance.

Lemma 11.5. *There exists a linear and continuous mapping* $\mathsf{E} : \mathcal{S}_\mathrm{e} \to \mathcal{S}_\mathrm{r}$ *satisfying*

$$\mathbf{M}^*\, \mathsf{E} = 1_{\mathcal{S}_\mathrm{e}}\,, \tag{11.13}$$

where $1_{\mathcal{S}_\mathrm{e}}$ *denotes the identity on* $\mathcal{S}_\mathrm{e}$, $1_{\mathcal{S}_\mathrm{e}}(f) = f$, $f \in \mathcal{S}_\mathrm{e}$. *For* $\varrho \geq \|\sigma\|$ *the mapping* E *is characterized by the Fourier transform*

$$\mathbf{F}\,\mathsf{E}f(\sigma,\varrho) = \hat{f}(\sigma, \sqrt{\varrho^2 - \|\sigma\|^2})\,, \qquad \sigma \in \mathbb{R}^n\,,\ \varrho \geq 0\,. \tag{11.14}$$

Identity (11.13) can easily be deduced from (11.14) with the help of representation (11.9). The crucial difficulty of proving Lemma 11.5 is to extend

F E for $\varrho < \|\sigma\|$. ANDERSSON uses in [5] an extension theorem contained in the book of STEIN [123]. Since this theorem does not provide an explicit representation of **F** E, we will use another technique to obtain an extension in Chapter 12.

To increase the efficiency of the inversion method, we used the existence of invariances in case of an operator between Hilbert spaces. Lemma 4.5 promises an improvement in efficiency in the distributional case, too, as long as an intertwining property applies to the adjoint $\mathbf{M}^*$, since $\mathbf{M}$ is one-to-one on $\mathcal{S}_\mathrm{e}'$. Lemma 11.6 will show that such an intertwining in fact does exist. But first, we introduce some notations.

For real $M > 1$ we distinguish certain open subsets of $\mathbb{R}^{n+1}$. We define

$$\begin{aligned} \mathcal{H}^M &:= \mathcal{H}^M(\mathbb{R}^{n+1}) = \{y = (y', y_{n+1})^\top \in \mathbb{R}^{n+1} : 1/M < |y_{n+1}|\}, \\ \mathcal{H}^{M,M} &:= \mathcal{H}^{M,M}(\mathbb{R}^{n+1}) = \{y = (y', y_{n+1})^\top \in \mathbb{R}^{n+1} : 1/M < |y_{n+1}| < M\}. \end{aligned}$$

Since the invariances in (11.15), (11.16) use dilations in y_{n+1}, the reconstruction points y must be contained in the complement of the hyperplane $y_{n+1} = 0$. That is why we introduced the set $\mathcal{H}^M$. To state convergence results as in Corollary 11.7 we even have to postulate that y_{n+1} is bounded. That is the reason to define $\mathcal{H}^{M,M}$. For an open subset $U \subset \mathbb{R}^{n+1}$ we denote

$$\begin{aligned} \mathcal{S}_\mathrm{e}(U) &:= \{v \in \mathcal{S}_\mathrm{e} : \operatorname{supp} v \subset U\}, \\ \mathcal{S}_\mathrm{e}'(U) &:= \{\lambda \in \mathcal{S}_\mathrm{e}' : \operatorname{supp} \lambda \subset U\}, \\ \mathcal{E}_\mathrm{e}'(U) &:= \{\lambda \in \mathcal{S}_\mathrm{e}' : \operatorname{supp} \lambda \subset U \text{ is compact}\}. \end{aligned}$$

Note that in general $\mathcal{S}_\mathrm{e}'(U)$ represents a proper subspace of $\mathcal{S}_\mathrm{e}(U)'$. Finally, let mappings $\mathcal{T}^y_{\mathrm{e},M} : \mathcal{S}_\mathrm{e} \to \mathcal{S}_\mathrm{e}$ and $\mathcal{G}^y_{\mathrm{r},M} : \mathcal{S}_\mathrm{r} \to \mathcal{S}_\mathrm{r}$ be defined by

$$\mathcal{T}^y_{\mathrm{e},M} v(x) = \begin{cases} |y_{n+1}|^{-n-1} v\Big(\frac{x'-y'}{|y_{n+1}|}, \frac{x_{n+1}}{|y_{n+1}|}\Big), & y \in \mathcal{H}^M(\mathbb{R}^{n+1}), \\ 0, & y \notin \mathcal{H}^M(\mathbb{R}^{n+1}), \end{cases} \tag{11.15}$$

$$\mathcal{G}^y_{\mathrm{r},M} w(z,r) = \begin{cases} |y_{n+1}|^{-2n-1} w\Big(\frac{z-y'}{|y_{n+1}|}, \frac{r}{|y_{n+1}|}\Big), & y \in \mathcal{H}^M(\mathbb{R}^{n+1}), \\ 0, & y \notin \mathcal{H}^M(\mathbb{R}^{n+1}). \end{cases} \tag{11.16}$$

Obviously, $\mathcal{T}^y_{\mathrm{e},M}$ and $\mathcal{G}^y_{\mathrm{r},M}$ are linear and bounded as compositions of translations and dilations. But nevertheless $\mathcal{T}^y_{\mathrm{e},M} v$ as well as $\mathcal{G}^y_{\mathrm{r},M} w$ may be discontinuous in y for $y_{n+1} = \pm 1/M$. Both mappings fulfill the desired intertwining property with respect to $\mathbf{M}^*$.

Lemma 11.6. *Let $\mathcal{T}^y_{\mathrm{e},M} : \mathcal{S}_\mathrm{e} \to \mathcal{S}_\mathrm{e}$ and $\mathcal{G}^y_{\mathrm{r},M} : \mathcal{S}_\mathrm{r} \to \mathcal{S}_\mathrm{r}$ be given as in* (11.15), (11.16) *respectively. Then*

$$\mathcal{T}^y_{\mathrm{e},M} \mathbf{M}^* = \mathbf{M}^* \mathcal{G}^y_{\mathrm{r},M}, \qquad y \in \mathbb{R}^{n+1}. \tag{11.17}$$

Proof. When $y \notin \mathcal{H}^M(\mathbb{R}^{n+1})$, then there is nothing to show, since both sides of (11.17) are equal to zero.

Let $y \in \mathcal{H}^M(\mathbb{R}^{n+1})$. Using representation (11.8) and appropriate substitutions yield

$$\begin{aligned}
&\mathbf{M}^* \mathcal{G}^y_{\mathrm{r},M} w(x', x_{n+1}) = \\
&= |y_{n+1}|^{-2n-1} \int_{\mathbb{R}^n} w\Big(\frac{z-y'}{|y_{n+1}|}, |y_{n+1}|^{-1} \sqrt{\|z-x'\|^2 + x_{n+1}^2}\Big)\,\mathrm{d}z \\
&= |y_{n+1}|^{-n-1} \int_{\mathbb{R}^n} w\Big(z - |y_{n+1}|^{-1} y', \sqrt{\|z - |y_{n+1}|^{-1} x'\|^2 + |y_{n+1}|^{-2} x_{n+1}^2}\Big)\mathrm{d}z \\
&= |y_{n+1}|^{-n-1} \int_{\mathbb{R}^n} w\Big(z, \sqrt{\|z - |y_{n+1}|^{-1} (x'-y')\|^2 + |y_{n+1}|^{-2} x_{n+1}^2}\Big)\,\mathrm{d}z \\
&= \mathcal{T}^y_{\mathrm{e},M}\, \mathbf{M}^* w(x', x_{n+1}),
\end{aligned}$$

where $w \in \mathcal{S}_{\mathrm{r}}$. This completes the proof. □

Lemma 11.6 allows for solving equation (11.12) for a *single* $y \in \mathbb{R}^{n+1}$ only.

We conclude this section by remarking that the transform $\mathcal{T}^y_{\mathrm{e},\infty} := \lim_{M\to\infty} \mathcal{T}^y_{\mathrm{e},M}$ is a representation of the group $(\mathbb{R}^n, +) \times ((0, +\infty), \cdot)$. The identity element of that group is $(0, \dots, 0, 1)^\top$ which is exactly that point for which equation (11.12) is to be solved. Thus, the invariances are adjusted to the given measurement geometry.

11.3 Approximate inverse for M

We give an outline how to transfer the abstract framework of distributional approximate inverse from Chapter 4 to the spherical mean operator **M**. We identify $\mathbf{A} = \mathbf{M}$, $V = \mathcal{S}_{\mathrm{e}}$, $W = \mathcal{S}_{\mathrm{r}}$, $\mathcal{T}^y_1 = \mathcal{T}^y_{\mathrm{e},M}$ and $\mathcal{T}^y_2 = \mathcal{G}^y_{\mathrm{r},M}$ and consider first the continuous problem (11.11). At the end of this section, we briefly deal with the semi-discrete setting which is necessary for the implementation of the method in Chapter 14.

Assume that we have an $e_\gamma(y)$ at hand which is in $\mathcal{S}_{\mathrm{e}}$ for every $y \in \mathbb{R}^{n+1}$ and satisfies the requirements to be a mollifier according to Definition 4.1. The reconstruction kernel $v_\gamma(y)$ associated with $e_\gamma(y)$ solves equation (11.12) and is an element of $\mathcal{S}_{\mathrm{r}}$ for every $y \in \mathbb{R}^{n+1}$ due to part b) from Remark 4.3 and Theorem 11.3. Applying Lemma 11.5, we immediately see that

$$v_\gamma(y) = \mathsf{E} e_\gamma(y) \tag{11.18}$$

fulfills (11.12). The intertwining property (11.17) enables us to solve equation (11.18) for $y = (0, \dots, 0, 1) \in \mathbb{R}^{n+1}$ only.

Corollary 11.7. *For all $\gamma > 0$ let $\bar{e}_\gamma \in \mathcal{S}_{\mathrm{e}}(\mathbb{R}^{n+1})$ and $e_\gamma(y) \in \mathcal{S}_{\mathrm{e}}(\mathbb{R}^{n+1})$ be generated for fixed $M > 1$ by the transform $\mathcal{T}^y_{\mathrm{e},M}$,*

$$e_\gamma(y) = \mathcal{T}^y_{\mathrm{e},M} \bar{e}_\gamma\,, \qquad y \in \mathbb{R}^{n+1}. \tag{11.19}$$

Assume that e_γ is a mollifier for **M** *according to Definition* 4.1. *Then all corresponding reconstruction kernels $\upsilon_\gamma(y)$ are obtained by*

$$\bar{\upsilon}_\gamma = \mathsf{E}\bar{e}_\gamma \tag{11.20}$$

and

$$\upsilon_\gamma(y) = \upsilon_\gamma(y)(z,r) = \mathcal{G}^y_{\mathrm{r},M}\bar{\upsilon}_\gamma(z,r)\,. \tag{11.21}$$

If e_γ is a $(\mathcal{E}'_{\mathrm{e}}(\mathcal{H}^{M,M}), \mathcal{S}_{\mathrm{e}}(\mathcal{H}^{M,M}))$-mollifier according to Definition 4.1, *then*

$$\widetilde{\mathbf{M}}_\gamma\,\mathbf{M}f := \langle \mathbf{M}f, \upsilon_\gamma(\cdot)\rangle_{\mathcal{S}'_{\mathrm{r}}\times\mathcal{S}_{\mathrm{r}}} \to f \qquad \textit{as}\ \gamma \to 0$$

for all $f \in \mathcal{E}'_{\mathrm{e}}(\mathcal{H}^{M,M})$. That means

$$\lim_{\gamma\to 0}\langle\langle \mathbf{M}f, \upsilon_\gamma(\cdot)\rangle_{\mathcal{S}'_{\mathrm{r}}\times\mathcal{S}_{\mathrm{r}}}, \beta\rangle_{\mathcal{E}'_{\mathrm{e}}(\mathcal{H}^{M,M})\times\mathcal{S}_{\mathrm{e}}(\mathcal{H}^{M,M})} = \langle f, \beta\rangle_{\mathcal{E}'_{\mathrm{e}}(\mathcal{H}^{M,M})\times\mathcal{S}_{\mathrm{e}}(\mathcal{H}^{M,M})}$$

whenever $\beta \in \mathcal{S}_{\mathrm{e}}(\mathcal{H}^{M,M})$.

Proof. Obviously, $\bar{\upsilon} = \mathsf{E}\bar{e}_\gamma$ satisfies $\mathbf{M}^*\bar{\upsilon} = \bar{e}_\gamma$. Lemma 11.6 then gives the identities

$$e_\gamma(x,y) = \mathcal{T}^y_{\mathrm{e},M}\bar{e}_\gamma(x) = \mathcal{T}^y_{\mathrm{e},M}\,\mathbf{M}^*\bar{\upsilon}_\gamma = \mathbf{M}^*\,\mathcal{G}^y_{\mathrm{r},M}\bar{\upsilon}_\gamma(x) = \mathbf{M}^*\{\upsilon_\gamma(y)\}(x)\,.$$

Taking into account that $\mathbf{M}^*\upsilon_\gamma(y) = e_\gamma(y)$, the convergences are conclusions from Definition 4.1. □

Remark 11.8. The fact that $e_\gamma(y)$ is generated by $\mathcal{T}^y_{\mathrm{e},M}$ implies that

$$\operatorname{supp}\widetilde{\mathbf{M}}_\gamma\,\mathbf{M}f \subset \mathcal{H}^M\,.$$

As a consequence we can only recover objects $f(y)$ whose support has a distance greater than $1/M$ from the plane $\{y_{n+1} = 0\}$. This is not a restriction for applications in SONAR and SAR, since the objects to be detected always have a positive distance from the measure plane $\{y_{n+1} = 0\}$. Thus, the objects always are supported in $\mathcal{H}^M(\mathbb{R}^{n+1})$ for sufficiently large M.
We will present a criterion for $\bar{e}_\gamma$ which guarantees that (11.19) generates a $(\mathcal{E}'_{\mathrm{e}}(\mathcal{H}^{M,M}), \mathcal{S}_{\mathrm{e}}(\mathcal{H}^{M,M}))$-mollifier in Chapter 12. Essentially, it is sufficient for $\bar{e}_\gamma$ to have mean value 1. Note that Corollary 11.7 says that using a $(\mathcal{E}'_{\mathrm{e}}(\mathcal{H}^{M,M}), \mathcal{S}_{\mathrm{e}}(\mathcal{H}^{M,M}))$-mollifier we have (weak) convergence of $\widetilde{\mathbf{M}}_\gamma\,\mathbf{M}f$ for distributions f with support in $\mathcal{H}^{M,M}(\mathbb{R}^{n+1})$ only. But again, M may be arbitrarily large.

Besides the translation invariance **M** has a dilation invariance, too. We have

$$\mathbf{M}^*\,\mathcal{D}^\gamma g(x) = \gamma^{-n-1}\,\mathbf{M}^*g(\gamma^{-1}x)$$

with $\mathcal{D}^\gamma g(z,r) = \gamma^{-2n-1}g(z/\gamma, r/\gamma)$. Thus, it would preferable to transfer this property to the mollifier $e_\gamma(y)$ by $e_\gamma(x,y) := \gamma^{-n-1}\,\mathcal{T}^y_{\mathrm{e},M}\bar{e}_1(x/\gamma)$. But unfortunately, such an e_γ does not fulfill the mollifier property of Definition 4.1 anymore.

We summarize the method of approximate inverse for solving $\mathbf{M}f = g$, $f \in \mathcal{E}'_{\mathrm{e}}(\mathcal{H}^{M,M})$.

- Choose $\bar{e}_\gamma \in \mathcal{S}_{\mathrm{e}}(\mathbb{R}^{n+1})$ such that
$$e_\gamma(x,y) = \mathcal{T}^y_{\mathrm{e},M}\bar{e}_\gamma(x)$$
is a mollifier.
- Compute $\bar{v}_\gamma = \mathsf{E}\bar{e}_\gamma$.
- Evaluate
$$\widetilde{\mathbf{M}}_\gamma g(y) = \langle g, \mathcal{G}^y_{\mathrm{r},M}\bar{v}_\gamma\rangle_{\mathcal{S}'_{\mathrm{r}}\times\mathcal{S}_{\mathrm{r}}} \tag{11.22}$$
for $y \in \mathcal{H}^{M,M}(\mathbb{R}^{n+1})$.

The crucial task in applying this algorithm is the computation of $\bar{v}_\gamma = \mathsf{E}\bar{e}_\gamma$. Representation (11.14) of $\mathbf{F}\,\mathsf{E}g$ is valid only for $\varrho \geq \|\sigma\|$. To calculate $\bar{v}_\gamma$, we need $\mathbf{F}\,\mathsf{E}\bar{e}_\gamma$ *for all* $\varrho \geq 0$. This fact has to be taken into account when designing an appropriate mollifier.

Remark 11.9. By means of Parseval's identity $\widetilde{\mathbf{M}}_\gamma g$ can be expressed by

$$\widetilde{\mathbf{M}}_\gamma g(y) = (2\,\pi)^{-2n-1}\,\langle \mathbf{F}g, \mathbf{F}\,\mathcal{G}^y_{\mathrm{r},M}\bar{v}_\gamma\rangle_{\mathcal{S}'_{\mathrm{r}}\times\mathcal{S}_{\mathrm{r}}}\,.$$

From (11.6) we see that

$$\operatorname{supp}\mathbf{F}g = \operatorname{supp}\mathbf{F}\,\mathbf{M}f \subset \{(\sigma,\varrho)\in\mathbb{R}^n\times[0,\infty) : \varrho \geq \|\sigma\|\}\,.$$

Hence, it seems sufficient to have knowledge of $\bar{v}_\gamma$ for $\varrho \geq \|\sigma\|$ only. But then it would be necessary to calculate the Fourier transform of the measured data, which ought to be avoided for two reasons. A discrete Fourier transform would extend the data periodically, which are known in a bounded domain only, leading to artifacts. On the other hand, we would have to calculate a three-dimensional Fourier transform in the 2D case ($n = 1$) and a Fourier transform in five dimensions for the 3D case ($n = 2$) which would decrease efficiency of the method significantly, since we could not use the radial symmetry in the last $n+1$ variables of $\mathbf{M}f$.

We conclude the chapter by dealing with the semi-discrete setting which is of great importance from a practical point of view. To this end, let $f \in \mathcal{E}'_{\mathrm{e}}(\mathcal{H}^{M,M})$ be such that $\mathbf{M}f$ can be identified with a continuous function which does not need to be integrable. If the measured data $\mathbf{M}f$ are given for $p+1$ centers $z_k \in \mathbb{R}^n$, $k = 0,\ldots,p$ and for $q+1$ radii r_l, $r_0 < r_1 < \ldots < r_q$, then we have to solve

$$\Psi_N\,\mathbf{M}f = g_N\,, \qquad g_N \in \mathbb{R}^N\,, \quad N = (p+1)(q+1)\,. \tag{11.23}$$

The observation operator $\Psi_N : \mathcal{C}(\mathbb{R}^n \times \mathbb{R}^+_0) \to \mathbb{R}^N$ is defined by

$$(\Psi_N w)_{k,l} = w(z_k, r_l)\,, \quad k = 0,\ldots,p\,, \quad l = 0,\ldots,q\,.$$

As we do not have a rigorous convergence theory as in the case of Hilbert spaces, we have to define the semi-discrete approximate inverse in another way. Since we have only a finite number of data and $\mathbf{M}f$ is a continuous function, the dual pairing on the right-hand side of (11.22) is a double integral with a bounded domain of integration. This suggests the application of numerical integration leading to

$$\widetilde{\mathbf{M}}_{N,\gamma} g_N(y) := \langle g_N, \mathcal{Q}_N \, \Psi_N \, \mathcal{G}^y_{\mathrm{r},M} \bar{\upsilon}_\gamma \rangle_{\mathbb{R}^N} . \tag{11.24}$$

The weights from numerical integration are contained in the matrix $\mathcal{Q}_N \in \mathbb{R}^{N \times N}$. The continuity of $\mathbf{M}f(z,r)\, \mathcal{G}^y_{\mathrm{r},M} \bar{\upsilon}_\gamma(z,r)$ yields pointwise convergence

$$\lim_{N \to \infty} \widetilde{\mathbf{M}}_{N,\gamma} \Psi_N \, \mathbf{M}f(y) = \langle \mathbf{M}f, \mathcal{G}^y_{\mathrm{r},M}, \bar{\upsilon}_\gamma \rangle_{L^2(\mathrm{ch}_\infty \times [0, r_\infty))} , \tag{11.25}$$

where ch_∞ and r_∞ are defined via

$$\mathrm{ch}_\infty := \bigcup_{p=1}^{\infty} \mathrm{ch}\big(\{z_k\}_{k=0}^p\big) , \qquad r_\infty := \lim_{q \to \infty} r_q$$

and $\mathrm{ch}(\{z_k\})$ denotes the convex hull of the centers $\{z_k\}$, $k = 0, \ldots, p$, in $\mathbb{R}^n$. The right-hand side of (11.25) equals $\langle f, e_\gamma(y) \rangle_{\mathcal{S}'_\mathrm{e} \times \mathcal{S}_\mathrm{e}}$, if $\mathrm{ch}_\infty = \mathbb{R}^n$ and $r_\infty = +\infty$. For $\gamma \to 0$, we obtain then (weak) convergence to f.

It is an open question, whether this is possible or not, and how the three parameters $\gamma \to 0, p, q \to \infty$ must be coupled to get convergence as in Corollary 8.4.

12

Design of a mollifier

This chapter is dedicated to the development of a $(\mathcal{E}'_e(\mathcal{H}^{M,M}), \mathcal{S}_e(\mathcal{H}^{M,M}))$-mollifier $e_\gamma(y)$ which fulfills the requirements of Corollary 11.7 and hence is generated by $\mathcal{T}^y_{e,M}$. Hence, let

$$e_\gamma(x,y) = \mathcal{T}^y_{e,M}\bar{e}_\gamma(x) = \begin{cases} |y_{n+1}|^{-n-1}\,\bar{e}_\gamma\left(\frac{x'-y'}{|y_{n+1}|}, \frac{x_{n+1}}{|y_{n+1}|}\right), & y \in \mathcal{H}^M(\mathbb{R}^{n+1}), \\ 0, & y \notin \mathcal{H}^M(\mathbb{R}^{n+1}) \end{cases} \tag{12.1}$$

for $\bar{e}_\gamma \in \mathcal{S}_e(\mathbb{R}^{n+1})$, $x = (x', x_{n+1})^\top$, $y = (y', y_{n+1})^\top$. Equation (12.1) implies $e_\gamma(x,y) = 0$ for $y \notin \mathcal{H}^M(\mathbb{R}^{n+1})$ and

$$e_\gamma(x, y', -y_{n+1}) = e_\gamma(x, y', y_{n+1}), \tag{12.2}$$

what shows that $e_\gamma(x,\cdot)$ is even in y_{n+1} and thus $\langle f, e_\gamma(\cdot,y)\rangle_{\mathcal{E}'_e\times\mathcal{S}_e}$ is even in y_{n+1}, too. Since we need $\mathsf{E}\bar{e}_\gamma$ to compute the reconstruction kernel, it is important that the Fourier transform of $\bar{e}_\gamma$ can be easily calculated in view of representation (11.14). For this reason, we assume $\bar{e}_\gamma$ to have a tensor product structure

$$\bar{e}_\gamma(x) = \bar{e}^1_\gamma(x') \otimes \bar{e}^2_\gamma(x_{n+1}), \tag{12.3}$$

where $\bar{e}^1_\gamma \in \mathcal{S}(\mathbb{R}^n)$ and $\bar{e}^2_\gamma$ is an even function in $\mathcal{S}(\mathbb{R})$. Conditions (12.1), (12.3) imply that $e_\gamma(\cdot,y) \in \mathcal{S}_e(\mathbb{R}^{n+1})$ for all $y \in \mathbb{R}^{n+1}$.

To make sure that e_γ in fact is a mollifier and to get rid of the difficulties which come from extending $\mathbf{F}\,\mathsf{E}\bar{e}_\gamma(\sigma,\varrho)$ for $\varrho \geq \|\sigma\|$, we postulate $\bar{e}^1_\gamma$ and $\bar{e}^2_\gamma$ to satisfy

1. $\int_{\mathbb{R}^n} \bar{e}^1_\gamma(z)\,\mathrm{d}z = \int_{\mathbb{R}} \bar{e}^2_\gamma(t)\,\mathrm{d}t = 1$,

2. $\mathbf{F}\bar{e}^1_\gamma$ is easily to calculate,

3. $\mathbf{F}\bar{e}^2_\gamma(\sqrt{\xi})$ has a nice extension for $\xi < 0$,

see Schuster and Quinto [115, Section 4].

Theorem 12.1 will state that the first property suffices to turn e_γ into a mollifier. The requirements 2. and 3. have only the sense to facilitate the computation of the corresponding reconstruction kernels and to avoid the extension lemma from the book of STEIN [123]. This is why we will construct $\bar{e}^2_\gamma$ in such a way that the extension can easily be deduced.

We proceed as in Section 4 of [115]. By now, we omitted a specific choice of $\bar{e}^1_\gamma$ and $\bar{e}^2_\gamma$. Before proving the main theorem of this chapter, we have to be more precise. Let $\bar{e}^1_\gamma$ be generated by dilations of a function e^1

$$\bar{e}^1_\gamma(x') = \gamma^{-n}\, e^1(x'/\gamma) \quad \text{for } e^1 \in \mathcal{S}(\mathbb{R}^n) \text{ with } \int_{\mathbb{R}^n} e^1(z)\,\mathrm{d}z = 1\,. \tag{12.4}$$

Thus, $\bar{e}^1_\gamma$ is defined in the same way as in case of the Radon or Doppler transform, that means by spatial translations and a dilation in γ, see (2.3). We have to be more careful with respect to $\bar{e}^2_\gamma$. If $F \in \mathcal{S}_\mathrm{e}(\mathbb{R})$ is an even function with mean value equal to 1, then $F(\cdot/\gamma)$ is not an appropriate choice for $\bar{e}^2_\gamma$. Since $\mathcal{T}^y_{\mathrm{e},M}$ involves a dilation with respect to $|y_{n+1}|$, too, the mollifier property is violated when setting $\bar{e}^2_\gamma$ in such a way. Therefore we define $\bar{e}^2_\gamma$ via

$$\bar{e}^2_\gamma(t) = \frac{1}{2\gamma}\left\{F\Big(\frac{t+1}{\gamma}\Big) + F\Big(\frac{t-1}{\gamma}\Big)\right\} \text{ for } F \in \mathcal{S}_\mathrm{e}(\mathbb{R}) \text{ with } \int_{\mathbb{R}} F(t)\,\mathrm{d}t = 1. \tag{12.5}$$

Functions $\bar{e}^1_\gamma$ and $\bar{e}^2_\gamma$ defined by (12.4), (12.5) satisfy requirement 1. and $\bar{e}^2_\gamma$ is an even function. The desired extension property 3. is guaranteed, when there exists to $\bar{e}^2_\gamma$ a function g such that

$$\mathbf{F}\bar{e}^2_\gamma(\sqrt{\xi}) = g(\xi^2)\,. \tag{12.6}$$

Equation (12.6) is the starting point for the specification of F and hence of $\bar{e}^2_\gamma$.

Defining e_γ by (12.1), (12.3), (12.4) and (12.5) we get a mollifier. This is subject of Theorem 12.1. In case of L^2-spaces this fact is shown with the help of suitable substitutions and Lebesgue's dominated convergence theorem. In case of distributions the proof is more sophisticated, since we have to show the weak convergence $\lambda_\gamma \rightharpoonup \lambda$ for all distributions $\lambda \in \mathcal{E}'_\mathrm{e}(\mathcal{H}^{M,M})$.

Theorem 12.1. *Let $M > 1$ and functions $\bar{e}^1_\gamma$, $\bar{e}^2_\gamma$ be given by* (12.4) *and* (12.5). *Then $e_\gamma(y)$ defined via* (12.1), (12.3) *represents a $(\mathcal{E}'_\mathrm{e}(\mathcal{H}^{M,M}), \mathcal{S}_\mathrm{e}(\mathcal{H}^{M,M}))$-mollifier.*

Proof. We follow the lines of the proof of Theorem 4.1 in SCHUSTER AND QUINTO [115]. Let $M > 1$ be fixed. The proof consists of three parts. First, Lemma 12.2 shows that $\lambda_\gamma(y) := \langle\lambda, e_\gamma(\cdot, y)\rangle_{\mathcal{S}'_\mathrm{e}\times\mathcal{S}_\mathrm{e}}$ for every $\lambda \in \mathcal{S}'_\mathrm{e}$ is again a distribution in $\mathcal{S}'_\mathrm{e}$. After that, we prove a variant of Fubini's theorem for distributions (Lemma 12.3). The third part finally deals with the convergence $\lambda_\gamma \rightharpoonup \lambda$ in $\mathcal{E}'_\mathrm{e}(\mathcal{H}^{M,M}) \times \mathcal{S}_\mathrm{e}(\mathcal{H}^{M,M})$. To show this convergence, we need a further, technical Lemma 12.4.

Lemma 12.2. *Let $\gamma > 0$ be fixed, e_γ be defined by* (12.1), (12.3), (12.4), (12.5) *and $\lambda \in \mathcal{S}_e'(\mathbb{R}^{n+1})$. Then $\lambda_\gamma(y) := \langle \lambda, e_\gamma(\cdot, y)\rangle_{\mathcal{S}_e' \times \mathcal{S}_e}$ represents a continuous function of polynomial growth for $y \in \mathcal{H}^M(\mathbb{R}^{n+1})$ which is identical to 0 when $y \notin \mathcal{H}^M(\mathbb{R}^{n+1})$. We conclude that $\lambda_\gamma \in \mathcal{S}_e'(\mathbb{R}^{n+1})$.*

Proof. (of Lemma 12.2)
Obviously $y \mapsto e_\gamma(\cdot, y)$ represents a continuous mapping between $\mathcal{H}^M$ and $\mathcal{S}_e$. Hence, λ_γ is continuous when $y \in \mathcal{H}^M$ and 0 else.
To show the polynomial increase of λ_γ, we apply Theorem 11.1 which allows a confinement to functions. Let P_λ be the function associated to the tempered distribution λ according to (11.3). For $y \in \mathcal{H}^M$, we obtain

$$
\begin{aligned}
\lambda_\gamma(y) &:= \langle \lambda, e_\gamma(\cdot, y)\rangle_{\mathcal{S}_e' \times \mathcal{S}_e} = (-1)^{|\alpha|} \int_{\mathbb{R}^{n+1}} P_\lambda(x)\, D_x^\alpha e_\gamma(x, y)\, \mathrm{d}x \\
&= \frac{(-1)^{|\alpha|}}{2} (\gamma\, |y_{n+1}|)^{-n-1} \int_{\mathbb{R}^n} \int_{\mathbb{R}} P_\lambda(x)\, D_{x'}^{\alpha'} \Big\{ e^1\Big(\frac{x'-y'}{\gamma\, |y_{n+1}|}\Big)\Big\} \\
&\quad \times D_{x_{n+1}}^{\alpha_{n+1}} \Big\{ F\Big(\frac{x_{n+1}}{\gamma\, |y_{n+1}|} - \frac{1}{\gamma}\Big) + F\Big(\frac{x_{n+1}}{\gamma\, |y_{n+1}|} + \frac{1}{\gamma}\Big)\Big\}\, \mathrm{d}x_{n+1}\, \mathrm{d}x' \qquad (12.7) \\
&= \frac{1}{2(-\gamma\, |y_{n+1}|)^{|\alpha|}} \int_{\mathbb{R}^n} \int_{\mathbb{R}} \Big\{ \Big[P_\lambda(\gamma\, |y_{n+1}|\, z' + y', \gamma\, |y_{n+1}|\, z_{n+1} + |y_{n+1}|) + \\
&\qquad + P_\lambda(\gamma\, |y_{n+1}|\, z' + y', \gamma\, |y_{n+1}|\, z_{n+1} - |y_{n+1}|)\Big] \\
&\qquad \times D^{\alpha'} e^1(z')\, D^{\alpha_{n+1}} F(z_{n+1})\, \mathrm{d}z_{n+1}\, \mathrm{d}z' \Big\},
\end{aligned}
$$

where we used the substitutions $z' = (x' - y')/(\gamma\, |y_{n+1}|)$ and $z_{n+1} = (x_{n+1}/|y_{n+1}| \pm 1)/\gamma$ as well as the symmetry of F. Since P_λ has a polynomial increase there exist constants $C_\lambda > 0$ and $\kappa > 0$ such that

$$
|P_\lambda(x)| \le C_\lambda\, (1 + \|x\|^2)^\kappa, \qquad \|x\| \to \infty, \quad x \in \mathbb{R}^{n+1}. \qquad (12.8)
$$

Using (12.8) to estimate the integrand in (12.7) yields

$$
\begin{aligned}
&\Big| P_\lambda(\gamma\, |y_{n+1}|\, z' + y', \gamma\, |y_{n+1}|\, z_{n+1} \pm |y_{n+1}|)\Big| \\
&\qquad \le C_\lambda \Big(1 + \|\gamma\, |y_{n+1}|\, z' + y'\|^2 + \big(\gamma\, |y_{n+1}|\, z_{n+1} \pm |y_{n+1}|\big)^2\Big)^\kappa \\
&\qquad \le C_\lambda\, 2^\kappa\, (1 + \gamma^2\, |y_{n+1}|^2\, \|z\|^2)^\kappa\, (1 + \|y\|^2)^\kappa \\
&\qquad \le C_\lambda\, 2^\kappa\, (M^{-2} + \gamma^2\, \|z\|^2)^\kappa\, |y_{n+1}|^{2\kappa}\, (1 + \|y\|^2)^\kappa .
\end{aligned}
$$

Here, we made use of the triangle inequality and of the estimates $(1 + a + b) \le (1 + a)\,(1 + b)$ and $(a + b)^2 \le 2\,(a^2 + b^2)$ which are valid when $a, b \ge 0$. Finally, we get for (12.7) a bound

$$
|\lambda_\gamma(y)| \le C_\lambda\, 2^\kappa\, s_\gamma\, |y_{n+1}|^{2\kappa}\, (\gamma\, |y_{n+1}|)^{-|\alpha|}\, (1 + \|y\|^2)^\kappa, \qquad y \in \mathcal{H}^M
$$

with

$$s_\gamma := \int_{\mathbb{R}^n}\int_{\mathbb{R}} (M^{-2} + \gamma^2\,\|z\|^2)^\kappa\, D^{\alpha'} e^1(z')\, D^{\alpha_{n+1}} F(z_{n+1})\,\mathrm{d}z_{n+1}\,\mathrm{d}z' < \infty$$

finishing the proof of Lemma 12.2. □

The main idea for continuing the proof of Theorem 12.1 is as follows: to prove the convergence $\lambda_\gamma \rightharpoonup \lambda$, we would like to investigate the convergence of the dual pairing $\langle \lambda, \langle e_\gamma(x,\cdot), \beta\rangle\rangle_{\mathcal{S}_e'\times\mathcal{S}_e}$ for $\gamma \to 0$ rather than that of $\langle \lambda_\gamma, \beta\rangle_{\mathcal{S}_e'\times\mathcal{S}_e}$. In case of (measurable) functions, this can be done with the help of Fubini's theorem (see e.g RUDIN [106, Theorem 8.8]). In case of distributions, we again will use Theorem 11.1 to pull back to functions.

Lemma 12.3. *Let $\gamma > 0$ be fixed, e_γ be defined by (12.1), (12.3), (12.4), (12.5), $\lambda \in \mathcal{S}_e'(\mathbb{R}^{n+1})$ and $\beta \in \mathcal{S}_e(\mathbb{R}^{n+1})$ be an even, rapidly decreasing function. If furthermore*

$$\beta_\gamma(x) := \langle e_\gamma(x,\cdot), \beta\rangle_{\mathcal{S}_e'(\mathbb{R}^{n+1})\times\mathcal{S}_e(\mathcal{H}^M)}\,, \tag{12.9}$$

then we have $\beta_\gamma \in \mathcal{S}_e(\mathbb{R}^{n+1})$ and the intertwining

$$\langle \lambda_\gamma, \beta\rangle_{\mathcal{S}_e'(\mathbb{R}^{n+1})\times\mathcal{S}_e(\mathcal{H}^M)} = \langle \lambda, \beta_\gamma\rangle_{\mathcal{S}_e'(\mathbb{R}^{n+1})\times\mathcal{S}_e(\mathbb{R}^{n+1})}\,. \tag{12.10}$$

Note that β_γ is a function in x, whereas λ_γ varies in y.

Proof. (of Lemma 12.3)
We again use representation (12.7) and write

$$\langle \lambda_\gamma, \beta\rangle_{\mathcal{S}_e'(\mathbb{R}^{n+1})\times\mathcal{S}_e(\mathcal{H}^M)} = \int_{\mathcal{H}^M}\int_{\mathbb{R}^{n+1}} I_\lambda^\gamma(y,x)\,\mathrm{d}x\,\mathrm{d}y\,, \tag{12.11}$$

where

$$I_\lambda^\gamma(y',y_{n+1},x',x_{n+1}) := \frac{(-1)^{|\alpha|}}{2}\,(\gamma\,|y_{n+1}|)^{-n-1-|\alpha|}\,\beta(y',y_{n+1})\,P_\lambda(x',x_{n+1})$$
$$\times (D^{\alpha'}e^1)\Big(\frac{x'-y'}{\gamma\,|y_{n+1}|}\Big)\Big\{(D^{\alpha_{n+1}}F)\Big(\frac{x_{n+1}-|y_{n+1}|}{\gamma\,|y_{n+1}|}\Big) + (D^{\alpha_{n+1}}F)\Big(\frac{x_{n+1}+|y_{n+1}|}{\gamma\,|y_{n+1}|}\Big)\Big\}.$$

The application of (12.8), $y \in \mathcal{H}^M$, and the fact that F, β, e^1 are functions in $\mathcal{S}_e$ as well as the estimates

$$1 + (a/b)^2 \geq (1+a^2)/(1+b^2)\,, \qquad a \in \mathbb{R}\,,\quad b \in \mathbb{R}\backslash\{0\}\,,$$
$$(1 + \|a-b\|^2)^{-k} \leq 2^k\,(1+\|b\|^2)^k\,(1+\|a\|^2)^{-k}\,, \qquad a,b \in \mathbb{R}^n\,,\quad k \in \mathbb{N}\,,$$

leads to

$$
\begin{aligned}
&|I_\lambda^\gamma(y', y_{n+1}, x', x_{n+1})| \\
&\leq (C_\lambda/2)\,(1+\|x\|^2)^\kappa\,(\gamma/M)^{-n-1-|\alpha|}\,|\beta(y', y_{n+1})|\left(1+\frac{\|x'-y'\|^2}{\gamma^2\,y_{n+1}^2}\right)^{-\mu_1} \\
&\qquad\times\left\{\left(1+\frac{(x_{n+1}-|y_{n+1}|)^2}{\gamma^2\,y_{n+1}^2}\right)^{-\mu_2}+\left(1+\frac{(x_{n+1}+|y_{n+1}|)^2}{\gamma^2\,y_{n+1}^2}\right)^{-\mu_2}\right\} \\
&\leq \frac{C_\lambda}{2}\,\frac{(1+\|x\|^2)^\kappa}{(1+\|y\|^2)^{\mu_3}}\left(\frac{\gamma}{M}\right)^{-n-1-|\alpha|} \\
&\qquad\times\frac{(1+\|y'\|^2)^{\mu_1}}{(1+\|x'\|^2)^{\mu_1}}\,\frac{(1+|y_{n+1}|^2)^{\mu_2}}{(1+|x_{n+1}|^2)^{\mu_2}}\,[2(1+\gamma^2\,y_{n+1}^2)]^{\mu_1+\mu_2}
\end{aligned}
$$

for any $\mu_j \in \mathbb{N}$, $j = 1, 2, 3$. The latter estimate implies that for μ_1, μ_2, μ_3 sufficiently large, I_λ^γ in (12.11) is dominated by a function which is integrable in $(x, y) \in \mathbb{R}^{n+1} \times \mathcal{H}^M$. Hence, the classical Fubini theorem applies to $\langle \lambda_\gamma, \beta \rangle_{\mathcal{S}_e' \times \mathcal{S}_e}$ and we may change the order of integration in (12.11). Since the integrand is arbitrarily smooth for $y \in \mathcal{H}^M$, we further may pull the differential operator D^α in front of the inner integral. Representation (11.3) then proves (12.10).
It remains to show that β_γ from (12.9) actually lies in $\mathcal{S}_e(\mathbb{R}^{n+1})$. To this end, let $\alpha \in \mathbb{N}_0^{n+1}$ be an arbitrary multiindex. We will prove that $D^\alpha \beta_\gamma(x)$ decreases more rapidly than any polynomial as $\|x\| \to \infty$. From (12.9) we deduce

$$
\begin{aligned}
D^\alpha \beta_\gamma(x) &= \int_{y\in\mathcal{H}^M} \frac{(-1)^{|\alpha|}}{2}\,|y_{n+1}\gamma|^{-n-1-|\alpha|}\,(D^{\alpha'} e^1)((x'-y')/(|y_{n+1}|\gamma)) \\
&\quad\times\left\{(D^{\alpha_{n+1}}F)\Big(\big(\frac{x_{n+1}}{y_{n+1}}-1\big)/\gamma\Big)+(D^{\alpha_{n+1}}F)\Big(\big(\frac{x_{n+1}}{y_{n+1}}+1\big)/\gamma\Big)\right\}\beta(y)\,\mathrm{d}y\,.
\end{aligned}
$$

We find a constant $\tilde{c}_\gamma > 0$ such that

$$
|D^\alpha \beta_\gamma(y)| \leq \tilde{c}_\gamma\,(1+\|y'\|^2)^{-\mu_1}\,(1+y_{n+1}^2)^{-\mu_2}\,, \qquad (y', y_{n+1}) \in \mathbb{R}^{n+1}
$$

holds true for any integers $\mu_1, \mu_2 \in \mathbb{N}$, since $\beta \in \mathcal{S}_e(\mathcal{H}^M)$ and $\gamma > 0$ is fixed. Applying similar arguments as in the estimate of $|I_\lambda^\gamma|$ it finally follows that $\beta_\gamma \in \mathcal{S}_e(\mathbb{R}^{n+1})$ and the proof of Lemma 12.3 is complete. □

Lemma 12.3 helps us to relocate the investigations of convergence from $\lambda_\gamma \rightharpoonup \lambda$ to $\beta_\gamma \to \beta$ in $\mathcal{S}_e(\mathcal{H}^{M,M})$. If β_γ tends to β in $\mathcal{S}_e(\mathcal{H}^{M,M})$, then this would be equivalent to the convergence $\lambda_\gamma \rightharpoonup \lambda$ in $\mathcal{E}_e'(\mathcal{H}^{M,M})$ because of (12.10). The following lemma thus plays a key role in the proof of Theorem 12.1.

Lemma 12.4. *We again assume e_γ to be defined by* (12.1), (12.3), (12.4), (12.5), *$\beta \in \mathcal{S}_e(\mathcal{H}^{M,M})$, $\alpha \in \mathbb{N}_0^{n+1}$ to be a multiindex and β_γ as in* (12.9). *Then $D^\alpha \beta_\gamma(x) \to D^\alpha \beta(x)$ pointwise in $\mathcal{H}^{M,M}$ as $\gamma \to 0$ and $D^\alpha \beta_\gamma$ is uniformly bounded in $(x, \gamma) \in \mathcal{H}^{M,M} \times (0, 1)$.*

Proof. (of Lemma 12.4)
We use again the symmetry of F and write

$$\beta_\gamma(x) = \int_{\mathcal{H}^{M,M}} \frac{1}{(\gamma\,|y_{n+1}|)^{n+1}}\, e^1\!\left(\frac{x'-y'}{\gamma\,|y_{n+1}|}\right) F\!\left(\left(\frac{x_{n+1}}{y_{n+1}}-1\right)/\gamma\right) \beta(y)\,\mathrm{d}y_{n+1}\,\mathrm{d}y'. \tag{12.12}$$

Provided that $(x,y) \in \mathcal{H}^{M,M} \times \mathcal{H}^{M,M}$, we perform the substitutions

$$z' = (x'-y')/(|y_{n+1}|\gamma)\,, \qquad z_{n+1} = \left(\frac{x_{n+1}}{y_{n+1}} - 1\right)/\gamma\,. \tag{12.13}$$

We have

$$\frac{y_{n+1}}{x_{n+1}} = \frac{1}{\gamma z_{n+1}+1} \qquad \text{and} \qquad \frac{y_{n+1}\,dz_{n+1}}{x_{n+1}} = \frac{dy_{n+1}}{y_{n+1}\gamma}$$

and the domain of integration with respect to z_{n+1} satisfies

$$\frac{1}{M^2} < \frac{1}{M|x_{n+1}|} < \frac{1}{|\gamma z_{n+1}+1|} < \frac{M}{|x_{n+1}|} < M^2\,. \tag{12.14}$$

The integral in (12.12) then turns into

$$\begin{aligned}\beta_\gamma(x) = \int_{\mathbb{R}^n}\int_{1/|\gamma z_{n+1}+1|<M^2} & e^1(z')\,F(z_{n+1}) \times \\ \times\beta\left(x' - \frac{\gamma|x_{n+1}|}{|\gamma z_{n+1}+1|}z', \frac{x_{n+1}}{\gamma z_{n+1}+1}\right) & \frac{1}{|\gamma z_{n+1}+1|}\,\mathrm{d}z'\mathrm{d}z_{n+1}\,,\end{aligned} \tag{12.15}$$

where the integration limits of the inner integral are bounded independently from x_{n+1} since $1/M < |y_{n+1}| < M$ and the support of β is contained in $\mathbb{R}^n \times [1/M, M]$.
We aim to subtract $\beta(x)$ inside of the integral (12.15) and then letting $\gamma \to 0$. At first, we set

$$b_\gamma(x) = \beta(x)\int_{\mathbb{R}^n}\int_{1/|\gamma z_{n+1}+1|<M^2} e^1(z')\,F(z_{n+1})\,\frac{1}{|\gamma z_{n+1}+1|}\,\mathrm{d}z_{n+1}\,\mathrm{d}z'$$

and prove that $\beta_\gamma - b_\gamma$ tends to 0 in $\mathcal{S}_\mathrm{e}(\mathcal{H}^{M,M})$. Considering the derivative $D^\alpha[\beta_\gamma - b_\gamma]$, we get

$$\begin{aligned}D^\alpha[\beta_\gamma(x) - b_\gamma(x)] = \int_{\mathbb{R}^n}\int_{1/|\gamma z_{n+1}+1|<M^2} & e^1(z')\,F(z_{n+1}) \\ \times D^\alpha_x\left\{\beta\left(x' - \frac{\gamma|x_{n+1}|}{|\gamma z_{n+1}+1|}z', \frac{x_{n+1}}{\gamma z_{n+1}+1}\right) - \beta(x)\right\} & \frac{1}{|\gamma z_{n+1}+1|}\,\mathrm{d}z'\,\mathrm{d}z_{n+1}\,.\end{aligned} \tag{12.16}$$

For applying Lebesgue's dominated convergence theorem to (12.16), we have to show that the integrand is dominated by an integrable function for every

$x \in \mathcal{H}^{M,M}$ uniformly in $\gamma \in (0,1)$. To this end, we explicitly calculate the derivative which appears in the right-hande side of (12.16).
The differential operator D_x^α can be split into $D_x^\alpha = D_{x_{n+1}}^{\alpha_{n+1}} D_{x'}^{\alpha'}$. The differentiation with respect to x' can be easily handled since x' occurs in the first argument of β only. We may compute

$$D_x^\alpha \Big\{ \beta\Big(x' - \frac{\gamma\,|x_{n+1}|}{|\gamma\, z_{n+1}+1|}\, z', \frac{x_{n+1}}{\gamma\, z_{n+1}+1}\Big) - \beta(x) \Big\} = \tag{12.17}$$

$$D_{x_{n+1}}^{\alpha_{n+1}} \Big\{ (D_{x'}^{\alpha'} \beta)\Big(x' - \frac{\gamma\,|x_{n+1}|}{|\gamma\, z_{n+1}+1|}\, z', \frac{x_{n+1}}{\gamma z_{n+1}+1}\Big)\Big\} - D_x^\alpha \beta(x).$$

The differentiation with respect to x_{n+1} addresses both arguments of $D_{x'}^{\alpha'}\beta$. If $\delta' = (\delta_1, \ldots, \delta_n) \in \mathbb{N}_0^n$ denotes a multiindex, then we prove for $x_{n+1} > 1/M > 0$ the following representation of the integrand in (12.16) by means of complete induction:

$$\frac{e^1(z')\, F(z_{n+1})}{|\gamma\, z_{n+1}+1|} \Bigg[\sum_{0<|\delta'|\le \alpha_{n+1}} \Bigg\{ \frac{\gamma^{|\delta'|}\, (-z)^{\delta'}}{|\gamma\, z_{n+1}+1|^{|\delta'|}\, (\gamma\, z_{n+1}+1)^{\alpha_{n+1}-|\delta'|}}$$

$$\times \Big(D_{x_{n+1}}^{\alpha_{n+1}-|\delta'|}\, D_{x'}^{\delta'+\alpha'} \beta \Big)\Big(x' - \frac{\gamma\,|x_{n+1}|}{|\gamma\, z_{n+1}+1|}\, z', \frac{x_{n+1}}{\gamma\, z_{n+1}+1}\Big) \Bigg\} \tag{12.18}$$

$$+ (\gamma\, z_{n+1}+1)^{-\alpha_{n+1}}\, (D^\alpha \beta)\Big(x' - \frac{\gamma\,|x_{n+1}|}{|\gamma\, z_{n+1}+1|}\, z', \frac{x_{n+1}}{\gamma\, z_{n+1}+1}\Big) - D^\alpha \beta(x) \Bigg].$$

A similar formula can be obtained for $x_{n+1} < -1/M < 0$.
From (12.14) we read that $1/|\gamma\, z_{n+1}+1| < M^2$. From (12.18) we then conclude that the integrand in (12.16) is bounded above by an integrable function uniformly in $\gamma \in (0,1)$. Lebesgue's dominated convergence theorem gives that $D^\alpha[\beta_\gamma - b_\gamma] \to 0$ pointwise in $\mathcal{S}_e(\mathcal{H}^{M,M})$ as $\gamma \to 0$. Note that the terms appearing as arguments in the sum of (12.18) are multiples of γ.
The convergence $b_\gamma \to \beta$ in $\mathcal{S}_e(\mathcal{H}^{M,M})$ is obvious since e^1 and F have mean value 1. Thus, we may summarize

$$\lim_{\gamma\to 0} (\beta_\gamma - \beta) = \lim_{\gamma\to 0} (\beta_\gamma - b_\gamma) + \lim_{\gamma\to 0} (b_\gamma - \beta) = 0$$

in $\mathcal{S}_e(\mathcal{H}^{M,M})$.
The uniform boundedness of $D^\alpha \beta_\gamma$ in $\gamma \in (0,1)$ is shown using similar arguments. □

We are now able to complete the proof of Theorem 12.1. We remember that $\lambda \in \mathcal{E}_e'(\mathcal{H}^{M,M})$ is assumed to have compact support in $\mathcal{H}^{M,M}$. According to Theorem 11.1, we have $\lambda = D^\alpha P_\lambda$ with a continuous, slowly increasing function P_λ. Unfortunately, in contrast to λ, the function P_λ does not need to have compact support. Hence, we define a cut-off function $\psi(x) = \psi_1(x')\, \psi_2(x_{n+1})$

from $\mathcal{S}_e(\mathbb{R}^{n+1})$ which is supposed to be identically 1 on supp λ and compactly supported. More specifically, we assume that the support of ψ_2 is contained in $[-2M, -1/2M] \cup [1/2M, 2M]$. Then we have $\lambda = \psi\, D^\alpha P_\lambda$ and from Lemma 12.3 it follows that

$$\begin{aligned}\langle \varphi_\gamma, \beta \rangle_{\mathcal{S}'_e(\mathbb{R}^{n+1}) \times \mathcal{S}_e(\mathcal{H}^{M,M})} &= \langle \varphi, \beta_\gamma \rangle_{\mathcal{S}'_e(\mathcal{H}^{M,M}) \times \mathcal{S}_e(\mathbb{R}^{n+1})} \\ &= (-1)^{|\alpha|} \int_{\mathcal{H}^{M,M}} P_\varphi(x)\, D^\alpha \{\psi(x)\, \beta_\gamma(x)\}\, dx\,.\end{aligned} \tag{12.19}$$

With the help of the product rule for derivatives and the convergence result from Lemma 12.4, we see that the derivative on the right-hand side of (12.19) converges pointwise in x on any compact set and is uniformly bounded. A last application of Lebesgue's dominated convergence theorem finally finishes the proof of Theorem 12.1. □

Remark 12.5. To get the estimates in the proof of Theorem 12.1, we had to postulate that the distributions λ which satisfy the mollifier property have compact support in $\mathcal{H}^{M,M}$. It is unclear which requirements have to be fulfilled that $\mathcal{T}^y_{e,M} \bar{e}_\gamma$ generates a $(\mathcal{S}'_e(\mathbb{R}^{n+1}), \mathcal{S}_e(\mathbb{R}^{n+1}))$-mollifier.

To conclude this chapter, we present two functions $\bar{e}_\gamma$ satisfying the conditions of Theorem 12.1. First, we set

$$\bar{e}^1_\gamma(x') = \gamma^{-n}\, e^1(x'/\gamma), \quad e^1(x') = (2\,\pi)^{-n/2} \exp(-\|x'\|^2/2)\,, \quad x' \in \mathbb{R}^n\,. \tag{12.20}$$

Hence, $\bar{e}^1_\gamma$ has a structure as in (12.4), is a function in $\mathcal{S}(\mathbb{R}^n)$ and has mean value $\int_{\mathbb{R}^n} \bar{e}^1_\gamma(x')\, dx' = 1$. The function F defining $\bar{e}^2_\gamma$ via (12.5) is supposed to be defined as

$$F(t) := 2\,\mathbf{F}^{-1}\{\exp(-|\xi|^4)\}(2t)\,. \tag{12.21}$$

Obviously, this function satisfies relation (12.6) with $g(\xi) = \exp(-|\xi|^2)$. Furthermore, we have $F \in \mathcal{S}_e(\mathbb{R})$ and $\int_{\mathbb{R}} F(t)\, dt = \hat{F}(0) = 1$.

The Fourier transform of $\bar{e}_\gamma = \bar{e}^1_\gamma \otimes \bar{e}^2_\gamma$ is not compactly supported when we choose $\bar{e}^1_\gamma$ and $\bar{e}^2_\gamma$ as in (12.20), (12.21). For this reason we consider a further mollifier. Let

$$h(\xi) = \begin{cases} e^{1 - \frac{1}{1-|\xi|^2}}, & |\xi| < 1\,, \\ 0, & |\xi| \geq 1 \end{cases} \tag{12.22}$$

and

$$e^1(x') = \mathbf{F}^{-1}\{h(\|\cdot\|)\}(\|x'\|)\,, \qquad F(t) = \mathbf{F}^{-1}\{h(\xi^2)\}(t)\,. \tag{12.23}$$

The functions $\bar{e}^1_\gamma$ and $\bar{e}^2_\gamma$ again are defined by (12.4), (12.5), respectively. With the help of Theorem 12.1, the mollifier property of both settings for e_γ can easily be verified.

Corollary 12.6. *Let $M > 1$ and $\bar{e}_\gamma = \bar{e}^1_\gamma \otimes \bar{e}^2_\gamma$ be defined by* (12.20), (12.21), *or* (12.22), (12.23), *respectively. Then the requirements of Theorem* 12.1 *are satisfied and $e_\gamma(y) = \mathcal{T}^y_{e,M} \bar{e}_\gamma$ is a $(\mathcal{E}'_e(\mathcal{H}^{M,M}), \mathcal{S}_e(\mathcal{H}^{M,M}))$-mollifier.*

Figure 12.1 displays $\bar{e}_\gamma$ for $n = 1$ – the two-dimensional case – and $\gamma = 0.06$. The global maximum is attained at $(0,1)^\top$ which is the identity element of the group $(\mathbb{R},+) \times ((0,+\infty),\cdot)$ and the single point for which equation (11.12) is to be solved.

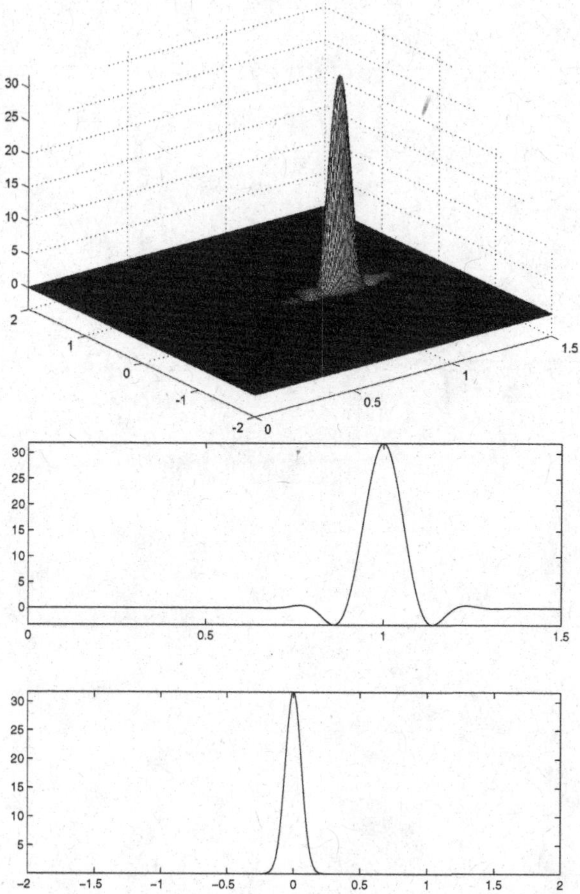

Fig. 12.1. Mollifier $\bar{e}_\gamma = \bar{e}^1_\gamma \otimes \bar{e}^2_\gamma$ defined by (12.5), (12.20) and (12.21) for $\gamma = 0.06$ (upper picture). Below, the graphs of $\bar{e}^1_\gamma$ (bottom) and $\bar{e}^2_\gamma$ (middle picture) are plotted. The width of the peak is about 0.5 units in each case (note the different scales), which is achieved by the dilation in (12.21). In contrast to the mollifiers considered in Part II, the function $\bar{e}_\gamma$ takes on negative values.

13

Computation of reconstruction kernels

This chapter is concerned with the computation of a reconstruction kernel associated with $\bar{e}_\gamma$, where the calculations are performed in detail for the mollifier given by (12.3), (12.5), (12.20) and (12.21). Our aim is to find a representation of

$$\bar{\upsilon}_\gamma = \mathsf{E}\bar{e}_\gamma \,. \tag{13.1}$$

The reconstruction kernel corresponding to $e_\gamma(x,y) = \mathcal{T}^y_{\mathrm{e},M}\bar{e}_\gamma(x)$ is then $\upsilon_\gamma(y) = \mathcal{G}^y_{\mathrm{r},M}\bar{\upsilon}_\gamma$ according to Corollary 11.7. From Lemma 11.5, we read that

$$\mathbf{F}\bar{\upsilon}_\gamma(\sigma,\varrho) = \mathbf{F}\,\mathsf{E}\bar{e}_\gamma(\sigma,\varrho) = \mathbf{F}\bar{e}_\gamma(\sigma,\sqrt{\varrho^2 - \|\sigma\|^2})\,, \tag{13.2}$$

when $\varrho \geq \|\sigma\|$, $\varrho \geq 0$ and $\sigma \in \mathbb{R}^n$. First, we have to calculate the Fourier transform of $\bar{e}_\gamma$.

Lemma 13.1. *We have*

$$\mathbf{F}\bar{e}_\gamma(\sigma,\varrho) = \hat{\bar{e}}_\gamma(\sigma,\varrho) = \hat{\bar{e}}^1_\gamma(\sigma)\,\hat{\bar{e}}^2_\gamma(\varrho) = \cos(\varrho)\,\mathrm{e}^{-\gamma^2\|\sigma\|^2/2}\,\mathrm{e}^{-\gamma^4\varrho^4} \tag{13.3}$$

for $\sigma \in \mathbb{R}^n$, $\varrho \in \mathbb{R}$.

Proof. Because of $\hat{\bar{e}}^1_\gamma(\sigma) = \mathrm{e}^{-\gamma^2\|\sigma\|^2/2}$ equation (13.3) follows from

$$\begin{aligned}
\hat{\bar{e}}^2_\gamma(\varrho) &= \frac{1}{2\gamma}\int_{\mathbb{R}}\left\{F\Big(\frac{q+1}{\gamma}\Big) + F\Big(\frac{q-1}{\gamma}\Big)\right\}\mathrm{e}^{-\imath q\varrho}\,\mathrm{d}q \\
&= \frac{1}{2}\left(\mathrm{e}^{\imath\varrho} + \mathrm{e}^{-\imath\varrho}\right)\int_{\mathbb{R}} F(q)\,\mathrm{e}^{-\imath\gamma q\varrho}\,\mathrm{d}q \\
&= \cos(\varrho)\,\mathrm{e}^{-(\gamma\varrho/2)^4}\,.
\end{aligned}$$

□

For $\varrho \geq \|\sigma\|$, we deduce

$$\mathbf{F}\bar{\upsilon}_\gamma(\sigma,\varrho) = \cos(\sqrt{\varrho^2 - \|\sigma\|^2})\,\mathrm{e}^{-\gamma^2\|\sigma\|^2/2}\,\mathrm{e}^{-\gamma^4(\varrho^2-\|\sigma\|^2)^2/16}\,. \tag{13.4}$$

from Lemma 13.1 and (13.2). With the help of (13.4), we see that an extension of $\hat{v}_\gamma$ to the whole of $\mathbb{R}^n \times [0,\infty)$ as a function from $\mathcal{S}_r$ requires an extension of $\cos(\sqrt{\xi})$ for $\xi < 0$. Thus, we need a function $G \in \mathcal{C}^\infty(\mathbb{R})$ with $G(\xi) = \cos(\sqrt{\xi})$, if $\xi \geq 0$, such that

$$\mathbf{F}\bar{v}_\gamma(\sigma,\varrho) = G(\varrho^2 - \|\sigma\|^2)\, \mathrm{e}^{-\gamma^2\|\sigma\|^2/2} \mathrm{e}^{-\gamma^4(\varrho^2-\|\sigma\|^2)^2/16} \tag{13.5}$$

is meaningfully defined for *all* $\sigma \in \mathbb{R}^n$, $\varrho \geq 0$ and additionally is a function in $\mathcal{S}_r$. The latter one implies that $\bar{v}_\gamma \in \mathcal{S}_r$.

The first idea to extend $\cos(\sqrt{\xi})$ is to use its power series expansion. For $\xi \geq 0$ we have

$$\cos(\sqrt{\xi}) = \sum_{k=0}^{\infty} \frac{(-1)^k}{2\,k!}\, \xi^k\,. \tag{13.6}$$

Using the power series (13.6) to extend $\cos(\sqrt{\xi})$ on $\xi < 0$ we obtain the function

$$G(\xi) = \begin{cases} \cos(\sqrt{\xi})\,, & \xi \geq 0\,, \\ \cosh(\sqrt{|\xi|})\,, & \xi < 0 \end{cases}$$

which obviously is in $\mathcal{C}^\infty(\mathbb{R})$, but unbounded. If we take into account that $G(\xi) = \mathcal{O}(\exp(\sqrt{|\xi|}))$ for $\xi \to -\infty$, then in fact we have that $\mathbf{F}\bar{v}_\gamma \in \mathcal{S}_r$.

As outlined in Remark 11.9 the particular choice of the extension for $\mathbf{F}\bar{v}_\gamma$ on $0 \leq \varrho < \|\sigma\|$ has no impact to $\widetilde{\mathbf{M}}_\gamma\, \mathbf{M}f$, since $\operatorname{supp} \mathbf{F}\,\mathbf{M}f \subset \{(\sigma,\varrho) : \varrho \geq \|\sigma\|\}$. In applications, we only have a finite number of data available as it was expressed by equation (11.23). This implies that the data are given on a bounded domain $(z,r) \in Z_N \times [0,R]$ only, where $Z_N \subset \mathbb{R}^n$ is bounded and $R > 0$. As a consequence, the specific extension of G actually has an influence to $\widetilde{\mathbf{M}}_{N,\gamma}\, \mathbf{M}f$. Numerical tests have shown that a bounded $G(\xi) \in \mathcal{C}(\mathbb{R})$ is desirable. To this end, we introduce a cut-off function $\chi \in \mathcal{C}^\infty(\mathbb{R})$ which is supposed to have the properties

$$\begin{aligned} \chi(\xi) &= 1\,, \quad \text{if } \xi \geq 0\,, \\ \chi(\xi) &= 0\,, \quad \text{if } \xi < -1\,, \\ \chi^{(k)}(-1) = \chi^{(k)}(0) &= 0\,, \quad \text{for all } k \geq 1\,. \end{aligned}$$

Such a function is explicitly given by

$$\chi(\xi) = \frac{u(\xi+1)}{u(\xi+1)+u(-\xi)}\,,$$

where

$$u(\xi) = \begin{cases} \mathrm{e}^{-1/\xi}\,, & \xi > 0\,, \\ 0\,, & \xi \leq 0 \end{cases}$$

The bounded extension $\widetilde{G}$ of $\cos(\sqrt{\xi})$ finally reads as

$$\widetilde{G}(\xi) = \begin{cases} \cos(\sqrt{\xi}), & \xi \geq 0, \\ \chi(\xi)\cosh(\sqrt{|\xi|}), & \xi < 0 \end{cases} \tag{13.7}$$

and is a bounded function in $\mathcal{C}^\infty(\mathbb{R})$. Plots of χ as well as of the extension $\widetilde{G}$ are displayed in Figure 13.1.

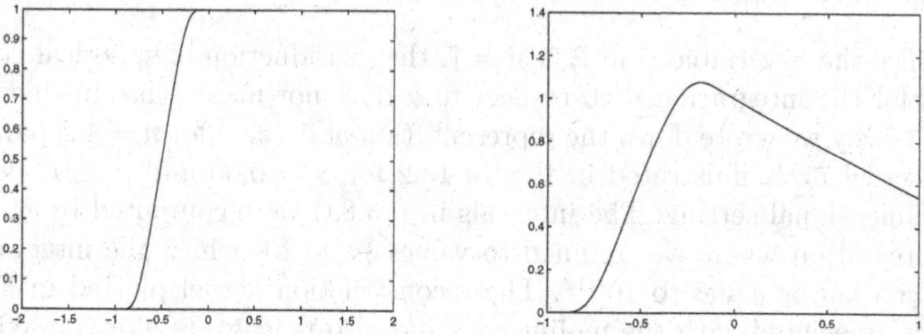

Fig. 13.1. Plots of the cut-off function χ (left picture) and the extension $\widetilde{G}$ (right picture). We have displayed $\widetilde{G}(\xi)$ only in the interval $\xi \in [-1, 1]$ to emphasize the smoothness of the extension.

The reconstruction kernel $\bar{v}_\gamma$ is now computed applying the inverse Fourier transform to (13.5).

Lemma 13.2. *Let* $\bar{e}_\gamma = \bar{e}_\gamma^1 \otimes \bar{e}_\gamma^2$ *be given by* (12.5), (12.20) *and* (12.21). *Then a solution of*

$$\mathbf{M}^* \bar{v}_\gamma = \bar{e}_\gamma$$

is represented by

$$\bar{v}_\gamma(z, r) = \frac{1}{2\pi^2} \int_0^\infty \int_0^\infty \Big\{ \widetilde{G}(\varrho^2 - \sigma^2)\, \mathrm{e}^{-\gamma^2 (\frac{\sigma^2}{2} + \gamma^2 (\varrho^2 - \sigma^2)^2/16)} \times \varrho\, \mathrm{J}_0(\varrho r) \cos(\sigma z) \Big\}\, \mathrm{d}\varrho\, \mathrm{d}\sigma \quad \textit{for } n = 1, \tag{13.8a}$$

$$\bar{v}_\gamma(z, r) = (2\pi)^{-n-\frac{1}{2}}\, r^{(1-n)/2}\, t^{(2-n)/2} \times \int_0^\infty \int_0^\infty \Big\{ \widetilde{G}(\varrho^2 - \tau^2)\, \mathrm{e}^{-\gamma^2 (\frac{\tau^2}{2} + \gamma^2 (\varrho^2 - \tau^2)^2/16)} \times \varrho^{(n+1)/2}\, \tau^{(2-n)/2}\, \mathrm{J}_{(n-1)/2}(\varrho r)\, \mathrm{J}_{(n-2)/2}(\tau t) \Big\}\, \mathrm{d}\varrho\, \mathrm{d}\tau \quad \textit{for } n > 1. \tag{13.8b}$$

Here $r, t > 0$, J_ν *denotes the Bessel function of first kind of order* ν *and* $\widetilde{G}$ *is defined as in* (13.7). *In* (13.8b) *we have* $t = \|z\|$ *and* $\tau = \|\sigma\|$.

Proof. Formulas (13.8a), (13.8b) follow from (13.5) by an application of the $(2n + 1)$-dimensional inverse Fourier transform and using identity (13.3) and spherical coordinates. The proof is completed with the help of

$$\int_{S^n} e^{\imath \varrho r \langle \omega,\theta \rangle} \, dS_n(\omega) = (2\pi)^{(n+1)/2} (\varrho r)^{(1-n)/2} J_{(n-1)/2}(\varrho r),$$

which is found e.g. in FAWCETT [30]. □

Since the σ-variable is in $\mathbb{R}$ if $n = 1$, the introduction of spherical coordinates for the integration with respect to σ does not make sense in that case. That is why we wrote down the representation of $\bar{v}_\gamma(z, r)$ for $n = 1$ separately. The kernel $\bar{v}_\gamma$ is illustrated in Figure 13.2 for $\gamma = 0.06$ and $n = 1$, i.e. the two-dimensional setting. The integrals in (13.8a) were computed by numerical integration where we confined to values (σ, ϱ) for which the integrand is greater than or equal to 10^{-12}. The reconstruction kernel plotted in Figure 13.2 is associated with the mollifier $\bar{e}_\gamma$ and also reaches its global maximum at $(0, 1)$. This again is compatible with the group structure which underlies the operators $\mathcal{G}^y_{r,M}$.

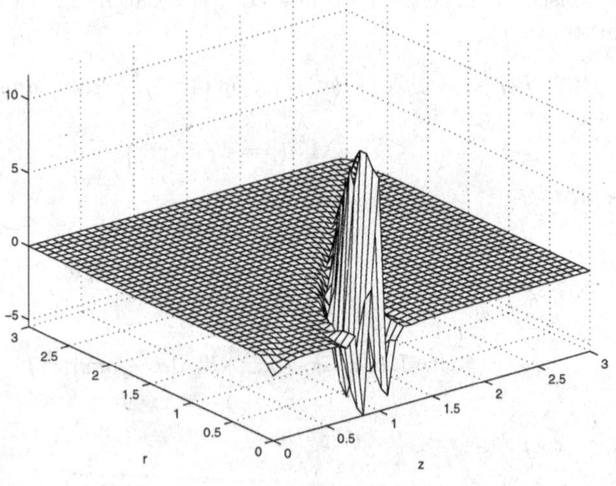

Fig. 13.2. The reconstruction kernel $\bar{v}_\gamma$ given as in (13.8a) for $\gamma = 0.06$ and $n = 1$. A cross section through the z-axis again would show the similarity to the Shepp-Logan filter just as in Doppler tomography, compare Figure 7.35, whereas we have a smoothing with respect to the radius variable r.

The computation of the reconstruction kernel corresponding to the mollifier with compactly supported Fourier transform is done accordingly.

Corollary 13.3. *If the mollifier* $\bar{e}_\gamma = \bar{e}_\gamma^1 \otimes \bar{e}_\gamma^2$ *is defined by* (12.4), (12.5), (12.22) *and* (12.23), *then a corresponding reconstruction kernel can be written as*

$$\bar{\upsilon}_\gamma(z,r) =$$

$$\frac{1}{2\pi^2}\Big\{\int_0^{\gamma^{-1}}\int_0^{\gamma^{-1}} \eta\, \mathrm{J}_0(\sqrt{\eta^2+\sigma^2}\,r)\,\cos\eta\, \mathrm{e}^{2-\frac{\gamma^2(\sigma^2-\gamma^2\eta^4)}{(1-\gamma^2\sigma^2)(1-\gamma^4\sigma^4)}}\,\cos(\sigma\, z)\,\mathrm{d}\eta\,\mathrm{d}\sigma$$

$$+\int_0^{\gamma^{-1}}\int_0^{\sigma} \eta\, \mathrm{J}_0(\sqrt{\sigma^2-\eta^2}\,r)\,\chi(-\eta^2)\,\cosh\eta\, \mathrm{e}^{2-\frac{\gamma^2(\sigma^2-\gamma^2\eta^4)}{(1-\gamma^2\sigma^2)(1-\gamma^4\sigma^4)}}\,\cos(\sigma\, z)\,\mathrm{d}\eta\,\mathrm{d}\sigma\Big\}$$

for $n = 1$, $z \in \mathbb{R}$, $r \geq 0$ *and*

$$\bar{\upsilon}_\gamma(z,r) = (2\pi)^{-n-\frac{1}{2}}\, r^{(1-n)/2}\, t^{(2-n)/2}$$

$$\times\Big\{\int_0^{\gamma^{-1}}\int_0^{\gamma^{-1}} \eta\, \mathrm{J}_{\frac{n-1}{2}}(\sqrt{\eta^2+\tau^2}\,r)\,\cos\eta\, \mathrm{e}^{2-\frac{\gamma^2(\tau^2-\gamma^2\eta^4)}{(1-\gamma^2\tau^2)(1-\gamma^4\tau^4)}}\,\mathrm{J}_{\frac{n-2}{2}}(\tau\, t)\,\mathrm{d}\eta\,\mathrm{d}\tau\, +$$

$$+\int_0^{\gamma^{-1}}\int_0^{\tau} \eta\, \mathrm{J}_{\frac{n-1}{2}}(\sqrt{\tau^2-\eta^2}\,r)\,\chi(-\eta^2)\,\cosh\eta\, \mathrm{e}^{2-\frac{\gamma^2(\tau^2-\gamma^2\eta^4)}{(1-\gamma^2\tau^2)(1-\gamma^4\tau^4)}}\,\mathrm{J}_{\frac{n-2}{2}}(\tau\, t)\,\mathrm{d}\eta\,\mathrm{d}\tau\Big\}$$

for $n > 1$, $t = \|z\|$, $\tau = \|\sigma\|$, *and* $r \geq 0$.

In Corollary 13.3, we additionally applied substitutions $\varrho = \sqrt{\eta^2+\sigma^2}$ and $\varrho = \sqrt{\sigma^2-\eta^2}$. Note that we do not need to restrict the integration limits in order to apply numerical integration since they are finite.

14

Numerical experiments

Over the last three chapters, we provided all tools for the implementation of the method of semi-discrete approximate inverse applied to $\Psi_N \mathbf{M} f$,

$$\widetilde{\mathbf{M}}_{N,\gamma} \Psi_N \mathbf{M} f(y) = \langle \Psi_N \mathbf{M} f, \mathcal{Q}_N \Psi_N \mathcal{G}^y_{\mathrm{r},M} \bar{v}_\gamma \rangle_{\mathbb{R}^N} . \tag{14.1}$$

We will test the performance of the method with the help of synthetic data in two dimensions ($n = 1$). Since $n = 1$, the measured data $\Psi_N \mathbf{M} f$ consist of a finite number of integrals over circles

$$\mathbf{M} f(z, r) = \frac{1}{2\pi} \int_{S^1} f(z + r\,\xi, r\,\eta)\, \mathrm{d}S_1(\xi, \eta)$$

$$= \frac{1}{2\pi} \int_0^{2\pi} f(z + r\,\cos\theta, r\,\sin\theta)\,\mathrm{d}\theta\,,$$

where $z \in \mathbb{R}$, $r \geq 0$. Applying the substitutions $\varrho = \sqrt{\eta^2 + \sigma^2}$ und $\varrho = \sqrt{\sigma^2 - \eta^2}$, respectively, the reconstruction kernel $\bar{v}_\gamma$ (13.8a) turns into

$$\bar{v}_\gamma(z, r) = \tag{14.2}$$

$$\frac{2}{(2\pi)^2} \Big\{ \int_0^\infty \int_0^\infty \eta\, \mathrm{J}_0(\sqrt{\eta^2 + \sigma^2}\, r)\, \cos\eta\, \mathrm{e}^{-\gamma^2 (\frac{\sigma^2}{2} + \frac{\gamma^2 \eta^4}{16})} \cos(\sigma z)\, \mathrm{d}\eta\, \mathrm{d}\sigma$$

$$+ \int_0^\infty \int_0^\sigma \eta\, \mathrm{J}_0(\sqrt{\sigma^2 - \eta^2}\, r)\, \chi(-\eta^2)\, \cosh\eta\, \mathrm{e}^{-\gamma^2 (\frac{\sigma^2}{2} + \frac{\gamma^2 \eta^4}{16})} \cos(\sigma z)\, \mathrm{d}\eta\, \mathrm{d}\sigma \Big\} .$$

Since we proved the mollifier property for e_γ for distributions with compact support in $\mathcal{H}^{M,M}(\mathbb{R}^n)$ only, see Theorem 12.1, we assume $f \in \mathcal{H}^{M,M}(\mathbb{R}^2)$ for a certain $M > 1$. The method of approximate inverse (11.22) for solving the continuous problem $\mathbf{M} f = g$ in two dimensions reads

$$\widetilde{\mathbf{M}}_\gamma \mathbf{M}f(y) = \langle \mathbf{M}f, \mathcal{G}^y_{\mathrm{r},M} \bar{\upsilon}_\gamma \rangle_{\mathcal{S}'_\mathrm{r} \times \mathcal{S}_\mathrm{r}} \tag{14.3}$$
$$= \frac{2\,\pi}{|y_2|^3} \int_{-\infty}^{\infty} \int_0^{\infty} r\, \mathbf{M}f(z,r)\, \bar{\upsilon}_\gamma\Big(\frac{z-y_1}{|y_2|}, \frac{r}{|y_2|}\Big)\, \mathrm{d}r\, \mathrm{d}z$$

with reconstruction points $y \in \mathcal{H}^{M,M}(\mathbb{R}^2)$, $y = (y_1, y_2)^\top$. Note that the right-hand side of (14.3) is not of filtered backprojection type due to the dilation by $|y_2|^{-1}$.

We show how the semi-discrete algorithm $\widetilde{\mathbf{M}}_{N,\gamma}$ emerges from (14.3). In practical situations, we only have a finite number of data $\mathbf{M}f(z,r)$ at hand. These are spherical means for finitely many centers $(z_k, 0)^\top$ with $z_k \in [\Lambda_1, \Lambda_2]$, $\Lambda_1 < \Lambda_2$, and radii $r_l \in [0, R]$ with $R > 0$. Assume that the sampling scheme is given as

$$z_k = \Lambda_1 + k \cdot h_z\,, \quad h_z = \frac{\Lambda_2 - \Lambda_1}{p}\,, \quad k = 0, \ldots, p\,,$$
$$r_l = l \cdot h_r\,, \quad h_r = \frac{R}{q}\,, \quad l = 0, \ldots, q\,. \tag{14.4}$$

Hence, we have $N = (p+1)\,(q+1)$ spherical means available. Instead of $\mathbf{M}f$, the measured data are given by the vector $\Psi_N\,\mathbf{M}f$, where the functionals $\Psi_N : \mathcal{C}(\mathbb{R} \times [0,\infty)) \to \mathbb{R}^N$ are point evaluations at (z_k, r_l)

$$(\Psi_N v)_{k,l} = v(z_k, r_l)\,, \quad 0 \le k \le p\,, \quad 0 \le l \le q\,, \quad v \in \mathcal{C}(\mathbb{R} \times [0,\infty))\,.$$

Remark 14.1. The observation operator Ψ_N is well-defined only if $\mathbf{M}f$ can be represented by a continuous function. Since we only know $\mathbf{M}f \in \mathcal{S}'_\mathrm{r}$, we have to require – just as in Section 11.3 – that $f \in \mathcal{E}'_\mathrm{e}(\mathcal{H}^{M,M}(\mathbb{R}^2))$ is such that $\mathbf{M}f \in \mathcal{C}(\mathbb{R} \times [0,\infty))$ holds. This does not mean a large restriction, since $\mathbf{M}$ smoothes by a factor of $1/2$ in Sobolev scales, see estimate (11.10). If, e.g., f is a sum of characteristic functions of convex sets with smooth boundaries, then actually $\mathbf{M}f$ is continuous. Note that we do not assume that $\mathbf{M}f$ is integrable over $\mathbb{R} \times [0,\infty)$ which in general does not hold true, see Figure 11.2.

To recover f from $\Psi_N\,\mathbf{M}f$, we apply the trapezoidal sum corresponding to the nodes $\{(z_k, r_l)\}$ to (14.3) and get

$$\widetilde{\mathbf{M}}_{N,\gamma}\, \Psi_N\, \mathbf{M}f = \langle \Psi_N\, \mathbf{M}f, \mathcal{Q}_N\, \Psi_N\, \mathcal{G}^y_{\mathrm{r},M} \bar{\upsilon}_\gamma \rangle_{\mathbb{R}^N} \tag{14.5}$$
$$= \frac{2\,\pi}{|y_2|^3}\, h_z\, h_r \sum_{k=0}^{p} \sum_{l=0}^{q} r_l\, \bar{\upsilon}_\gamma\Big(\frac{z_k - y_1}{|y_2|}, \frac{r_l}{|y_2|}\Big)\, \mathbf{M}f(z_k, r_l)$$

for $y \in \mathcal{H}^{M,M}(\mathbb{R}^2)$. In that case, we have $\mathcal{Q}_N = h_z\, h_r\, I_{N,N}$. The specific sampling scheme yields the convergence

$$\lim_{N\to\infty} \widetilde{\mathbf{M}}_{N,\gamma} \Psi_N \mathbf{M} f(y) = \langle \mathbf{M} f, \mathcal{G}^y_{\mathrm{r},M} \bar{\upsilon}_\gamma \rangle_{L^2([\Lambda_1,\Lambda_2]\times[0,R])} \,.$$

according to (11.25). Formula (14.5) was applied to get the reconstructions in figures 14.1 and 14.2.

The kernel $\bar{\upsilon}_\gamma$ has been pre-computed by applying numerical integration to (14.2) choosing convenient integration boundaries. Moreover, we determine $\bar{\upsilon}_\gamma(z,r)$ on the square $[0,10]^2$ on a equidistant mesh grid consisting of 128×128 grid points. Since the kernel is rapidly decreasing, the absolute value of $\bar{\Psi}_\gamma$ outside the square $[0,10]^2$ is rather small, so we may extend the kernel by 0 there. Using the symmetry $\bar{\upsilon}_\gamma(z,r) = \bar{\upsilon}_\gamma(-z,r)$ and linear interpolation, we get $\bar{\upsilon}_\gamma(z,r)$ for every $z \in \mathbb{R}$, $r \geq 0$. To check the performance of the above algorithm, we implemented it to reconstruct several objects. All reconstructions were computed for $(y_1,y_2) \in [0,7]\times[1,8]$ using an equidistant mesh grid with 128×128 grid points. The objects are assumed to have their support in $\mathcal{H}^1(\mathbb{R}^2)$. The data are given according to the sampling scheme (14.4) with $\Lambda_1 = -36$, $\Lambda_2 = 36$, $p = 384$, $R = 25$ and $q = 400$. Please note, that in all pictures the y_2-axis is the horizontal one, whereas the y_1-axis (the SONAR sources, circle centers) is the vertical one.

1.) First, we recovered the function χ_C (11.5), the characteristic function of two disks reflected about the y_1-axis. Hence, χ_C is even in y_2 and a tempered distribution yielding $\chi_C \in \mathcal{S}'_{\mathrm{e}}(\mathbb{R}^2)$. The picture at the top of Figure 14.1 shows the original function χ_C in $[0,7]\times[1,8]$ as well as the approximate inverse $\widetilde{\mathbf{M}}_{N,\gamma} \Psi_N \mathbf{M}\chi_C$. Cross sections of the reconstruction along with the exact object function χ_C are plotted in the bottom picture in Figure 14.1. The horizontal and vertical cross sections displayed in Figure 14.1 emphasize that the singularities of $f(y_1,y_2)$ for $y_1 = const$ can be better recovered than those for $y_2 = const$ as predicted by the microlocal analysis, see Remark 14.2.

2.) Second, we applied the algorithm to the sum of two functions where each one has a representation as χ_C and thus is in $\mathcal{S}'_{\mathrm{e}}(\mathbb{R}^2)$. The function to be recovered consists of four disks alltogether having radius 1. One disk is centered about $(4,4)$ and has density 2, the second is centered about $(2,3)$ with density 1 and both of them are reflected about the y_1-axis to obtain again a function which is even in y_2. The reconstruction, the original object as well as corresponding cross sections can be seen in Figure 14.2 for $(y_1,y_2) \in [0,7]\times[1,8]$, the parameters are the same as in Figure 14.1.

These tests show that the method of approximate inverse works fine, and the reconstructions would be comparable to those in DENISJUK [20] if the optimal regularization parameter was found (see Remark 14.2 for that problem). Some blurring in the reconstructions is caused by the numerical calculation of the reconstruction kernel and truncation errors. However, some ill-posedness is inherent in the problem.

Remark 14.2. Some of the fuzzy reconstruction boundaries in figures 14.1 and 14.2 are intrinsic to the problem. As shown in LOUIS, QUINTO [74], PALAMODOV [91], the object boundaries that are most difficult to reconstruct are

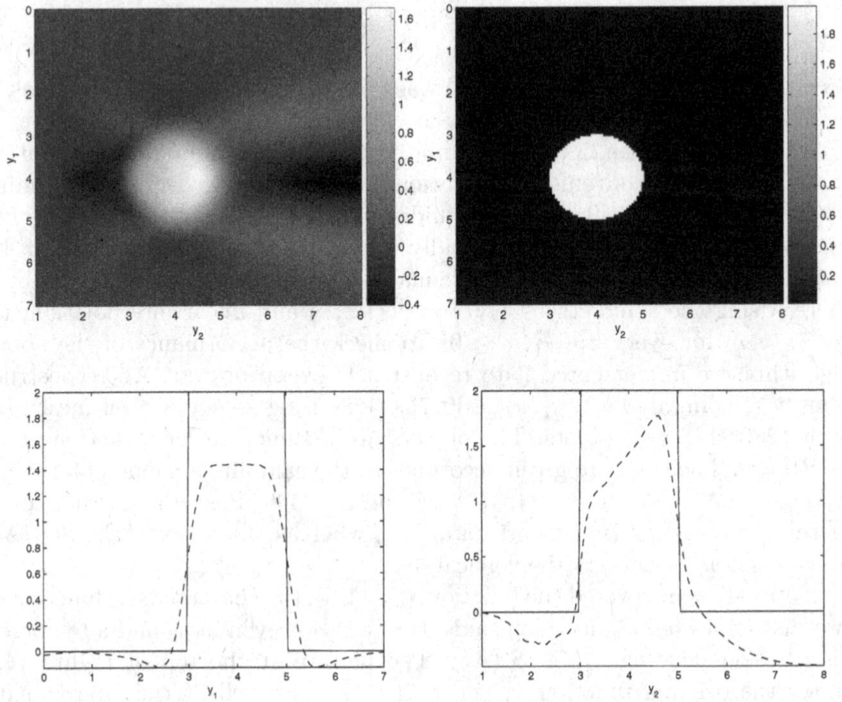

Fig. 14.1. Top: Reconstruction of χ_C as in (11.5), (left picture) and original object function (right picture), $\gamma = 0.04$. The y_1- and y_2-axes are switched. The sources are located at the vertical (y_1-) axis. Bottom: Cross sections through the reconstructions (dashed line) and the original object function χ_C (solid line). We have displayed $\widetilde{\mathbf{M}}_{N,\gamma}\, \Psi_N\, \mathbf{M}\chi_C(y_1, y_2)$ for $y_2 = 4$ (left picture) and for $y_1 = 4$ (right picture). One sees that the reconstruction is smooth and recovers the jumps of χ_C but the peak is attenuated. This is an indicator that we have not chosen the optimal regularization parameter γ which is difficult to determine. The jumps in the left-hand cross section furthermore belong to edges which are hard to recover according to the microlocal analysis, see Figure 14.3. This is the reason why the left jump is located more exactly in the right-hand picture.

those not tangent to circles in the data set. This means that horizontal boundaries in figures 14.1 and 14.2 will be intrinsically hardest to reconstruct, since the set of circle centers is the vertical axis. Since more-or-less vertical boundaries are tangent to spheres in the data set, the microlocal analysis predicts they will be easiest to reconstruct. This is analogous to limited angle X-ray tomography in which some boundaries are invisible in the data, see QUINTO [94]. We emphasized the situation in Figure 14.3.

We furthermore mention that the numerical computation of the reconstruction kernel breaks down for very small values of γ, since then the integrand in (14.2) has dramatic oscillations which are almost impossible to handle. That

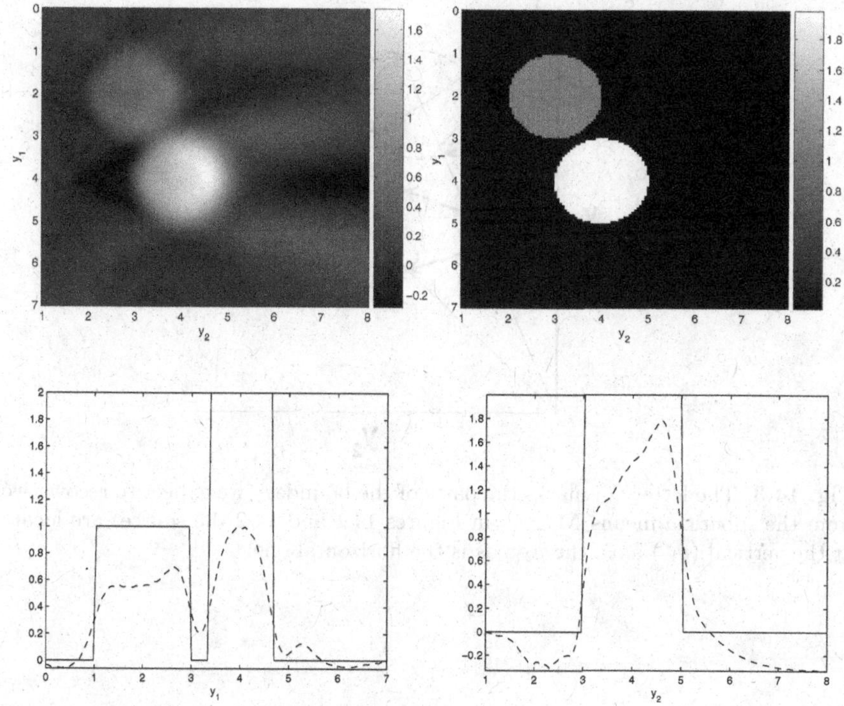

Fig. 14.2. Top: Reconstruction of the sum of two functions of type (11.5), (left picture) and original object function in $[0, 7] \times [1, 8]$ (right picture), $\gamma = 0.04$. The y_1- and y_2 axes are switched. The sources are located at the vertical (y_1-) axis. Bottom: Reconstruction profiles (dashed lines) $y_2 = 3.5$ (left picture) and $y_1 = 4$ (right picture) and corresponding profiles of the exact function (solid lines). Although the cross section for $y_2 = 3.5$ seems to be smoother, the location of the singularities are determined better in the profile $y_1 = 4$. This coincides with the moicrolocal analysis, see Remark 14.2.

makes it impossible to compute reconstructions with the optimal γ since the results in Figures 14.1, 14.2 suggest that the optimal γ is distinctly smaller than 0.04. Thus, analytic expressions for reconstruction kernels are of large interest.

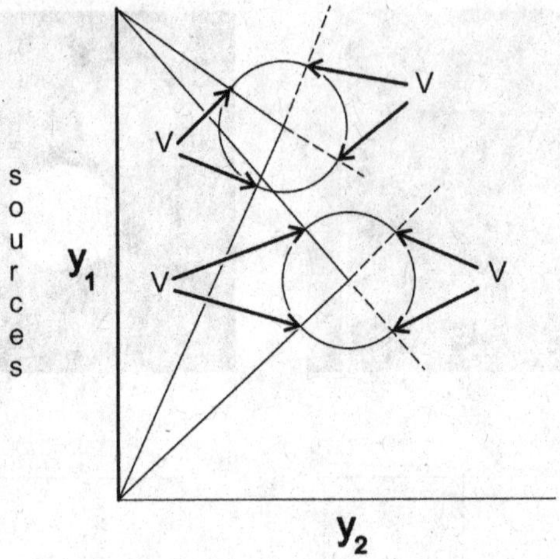

Fig. 14.3. The letter 'V' marks the parts of the boundary we expect to recover well from the spherical means $\mathbf{M}f$. As in Figures 14.1 and 14.2 the sources are located at the vertical (y_1-) axis, the y_2-axis is the horizontal one.

15

Conclusion and perspectives

The spherical mean operator **M** with hyperplanes as center sets is a typical example of a linear operator which cannot be formulated as a bounded mapping between Hilbert spaces. It has important applications in SONAR and SAR. The method of approximate inverse proved to be a powerful tool to construct an inversion scheme for this operator for the semi-discrete setting, too. Note that the computation of reconstruction kernels in three dimensions ($n = 2$) can be done with the same complexity as in the 2D case, since we use radial symmetric mollifiers, see (13.8b) and Corollary 13.3. Numerical experiments in 3D will be subject of further investigations. Questions concerning resolution of synthetic-aperture imaging have been answered by NATTERER ET AL. [83] and BORDEN AND CHENEY [10].

Current research deals furthermore with adapting ideas from local tomography (see FARIDANI ET AL. [29, 27], VAINBERG AND FAINGOIS [127]) to **M**. Instead of reconstructing the original object f, one might be interested in getting information about discontinuities of f, that means one reconstructs a function Λf having the same singular support as the original f. In case of **M** this could be done computing a pseudodifferential operator

$$\Lambda f := \Delta \mathbf{M}^* \Gamma_\varepsilon \mathbf{M} f ,$$

where Δ is the Laplacian and Γ_ε denotes the multiplication with a cut-off function which is necessary since the adjoint $\mathbf{M}^*$ can not be applied to **M** in the first place. The support of the cut-off function is controlled by a parameter ε. Certainly, the Laplacian Δ can be replaced by another appropriate elliptic pseudodifferential operator.

Part IV

Further Applications

In the last part of the book, we give brief overviews over some further applications where the method of approximate inverse has led to efficient solution schemes. Chapter 16 introduces to the problem of X-ray diffractometry, which is a method of non-destructive testing where one is interested in detecting the stress tensor of a given probe using X-ray measurements. Chapter 17 is concerned with the three-dimensional thermoacoustic computerized tomography (TCT). This is an inverse problem with applications in medical imaging and non-destructive testing, which is related to the SONAR problem investigated in Part III. Here, the center sets of the spherical mean operator are spheres and the approximate inverse leads to a stable inversion scheme of filtered backprojection type. In Section 2.2, we already dealt with the problem of 2D computerized tomography. In Chapter 18 we show how the method of approximate inverse can be used to compute reconstruction kernels in 3D computerized tomography when the X-ray sources lie on a curve.

16

Approximate inverse and X-ray diffractometry

16.1 X-ray diffractometry

The aim of X-ray diffractometry is to recover the stress tensor $\sigma = \sigma_{ij}$ in a probe, which has a crystalline structure, with the help of X-rays. A comprehensive treatise of this kind of non-destructive testing along with some solution schemes can be found in KÄMPFE [52]. We give here a short description of the measurement setting.

Mechanical stresses and elastic strains in solids are connected by *Hook's law*

$$\varepsilon_{ij} = \frac{\nu + 1}{E}\,\sigma_{ij} - \delta_{ij}\,\frac{\nu}{E}\sum_{k=1}^{3}\sigma_{kk}\,, \tag{16.1}$$

where ε_{ij} denotes the strain tensor, E is the modulus of elasticity and ν is the Poisson number. We introduce a laboratory coordinate system by rotating the probe about the x_3-axis by an angle φ and tilting the probe about the x_2-axis by an angle ψ which transforms the strain tensor to

$$\varepsilon_{ij}^{L} = \sum_{k,l=1}^{3} T_{ik}^{\varphi\psi}\,\varepsilon_{kl}\,T_{jl}^{\varphi\psi} \tag{16.2}$$

with

$$T^{\varphi\psi} = \begin{pmatrix} \cos\varphi\cos\psi & \sin\varphi\cos\psi & -\sin\psi \\ -\sin\varphi & \cos\varphi & 0 \\ \cos\varphi\sin\psi & \sin\varphi\sin\psi & \cos\psi \end{pmatrix}.$$

In X-ray diffractometry, it is only possible to measure the component

$$\varepsilon_{\varphi\psi} := \varepsilon_{33}^{L}\,.$$

Putting Hook's law (16.1) into (16.2) gives an explicit expression for $\varepsilon_{\varphi\psi}$ depending on the stress tensor σ_{ij}. To describe diffraction of X-rays at a crystal lattice, we use Bragg's reflection model, which explains deflections as reflections on parallel lattice planes, see Figure 16.1. According to *Bragg's condition*, the angle of the interference peak θ is related to the distance d of the lattice planes as

$$2\,d\,\sin\theta = n\,\lambda\,, \tag{16.3}$$

where λ means the wavelength of the applied X-rays and n is an integer.

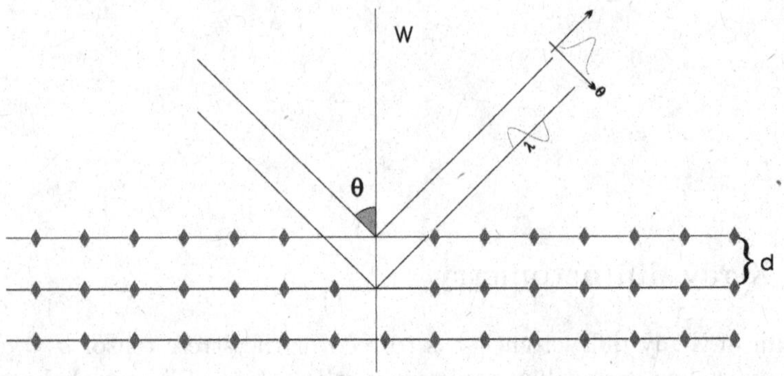

Fig. 16.1. Bragg's reflection model. W denotes the axis of rotation.

Stresses cause a shift of the peak position θ away from θ_0, the interference maximum of the unstressed probe. Differentiating (16.3) gives

$$\frac{\partial d}{\partial \theta} = -n\,\frac{\lambda\,\cos\theta}{2\,\sin^2\theta} = -d\,\cot\theta$$

and thus $\mathrm{d}d/d = -\cot\theta\,\mathrm{d}\theta$. A linearization finally identifies the measured strain component $\varepsilon_{\varphi,\psi}$ as a peak shift

$$\varepsilon_{\varphi\psi} = \frac{\mathrm{d}d}{d} \approx \frac{d_{\varphi\psi} - d_0}{d_0} = -\cot\theta_0\,(\theta_{\varphi\psi} - \theta_0)\,, \tag{16.4}$$

where $\theta_{\varphi\psi}$, $d_{\varphi\psi}$ means the peak position and lattice plane distance of the stressed probe corresponding to rotation angle φ and tilt angle ψ, d_0 is the distance of lattice planes in the unstressed specimen. Putting together (16.2), (16.4) and Hook's law (16.1) we get the formula

$$\varepsilon_{\varphi\psi} = -\cot\theta_0\,(\theta_{\varphi\psi} - \theta_0) = \sum_{k,l=1}^{3} T_{3k}^{\varphi\psi}\left(\frac{\nu+1}{E}\,\sigma_{kl} - \delta_{kl}\,\frac{\nu}{E}\sum_{m=1}^{3}\sigma_{mm}\right)T_{3l}^{\varphi\psi}\,, \tag{16.5}$$

which is valid for isotropic material and is the fundamental equation of X-ray based stress detection. The peak positions θ are also called *Bragg angles.*

So far, we did not take into account that the intensity $I(z)$ of the X-rays is attenuated within the probe according to *Lambert-Beer's law*

$$I(z) = I_0\,\mathrm{e}^{-\mu\,z}\,,$$

where I_0 is the intensity of the emitted rays, z denotes the penetration depth and μ is the attenuation coefficient. Hence, we obtain the fundamental equation of X-ray diffractometry from (16.5), when we replace the tensor entries σ_{ij} by their *reciprocal Laplace transforms* $\breve{\sigma}_{ij}$, where

$$\breve{\sigma}_{ij}(\tau_\psi) = \frac{1}{\tau_\psi} \int_0^\infty \sigma_{ij}(z)\, \mathrm{e}^{-z/\tau_\psi}\, \mathrm{d}z\,. \tag{16.6}$$

Here, $\tau_\psi = \cos\psi \, \sin\theta_0/(2\,\mu)$ denotes the maximal penetration depth of the X-ray beam, which depends on the tilt angle ψ. Thus, a method for X-ray diffractometry has to overcome two difficulties: decoupling the transformed tensor entries $\breve{\sigma}_{ij}$ from (16.5) followed by a solver for the reciprocal Laplace transform (16.6).

One of the most popular and oldest method for X-ray diffractometry is the $\sin^2\psi$-method, see MACHERAUCH, MÜLLER [76]. Here, one assumes the stresses to be near the surface and parallel to it, which leads to $\sigma_{13} = \sigma_{23} = \sigma_{33} = 0$. The basic formula then reads

$$\varepsilon_{\varphi\psi} = -\cot\theta_0\,(\theta_{\varphi\psi} - \theta_0) = \frac{\nu+1}{E}\,\sigma_\varphi\,\sin^2\psi - \frac{\nu}{E}\,(\sigma_{11}+\sigma_{22}). \tag{16.7}$$

with $\sigma_\varphi = \sigma_{11}\cos^2\varphi + \sigma_{12}\sin 2\varphi + \sigma_{22}\sin^2\varphi$. Equation (16.7) represents a line depending on $\sin^2\psi$. The $\sin^2\psi$-method recovers stress σ_φ, which acts in parallel to the surface, by identifying the slope of the best fit straight line from (16.7). Other methods rely on a least squares parameter fitting of certain functions (e.g. polynomials, splines) to the searched for stress tensor σ_{ij}, see e.g. EIGENMANN ET AL. [25], LEVERENZ ET AL. [63].

We describe a method published in SCHUSTER ET AL. [114] which gains the transformed tensor entries $\breve{\sigma}_{ij}$ from (16.5) applying a least squares method followed by inverting the reciprocal Laplace transforms with the help of the approximate inverse. The next section is concerned with this very step: applying the approximate inverse to the Laplace transform.

16.2 Approximate inverse for the Laplace transform

In this section we subsume results from SCHUSTER [111] where a stable inversion scheme for the Laplace transform

$$\mathbf{L}f(t) = \int_0^\infty f(z)\, \mathrm{e}^{-t\,z}\, \mathrm{d}z \tag{16.8}$$

was developed. We consider $\mathbf{L}$ as a mapping which acts on $L^2(\mathbb{R}^+)$. Thus, $\mathbf{L}$ is a linear, self-adjoint, but unbounded operator. In practical situations, $\mathbf{L}f$ is given only for finitely many scanning points t_j in an interval $[a, b]$, where

$t_j \neq t_i$, if $j \neq i$. In our setting the scanning points t_j do not need to be equidistant. E.g. in X-ray diffractometry, we have

$$t_j = 2\,\mu/(\sin\theta_0\,\cos\psi_j)\,. \tag{16.9}$$

That means we evaluate $\mathbf{L}$ at the points t_j and obtain the semi-discrete mapping

$$(\mathbf{L}_m f)_j = (\Psi_m\,\mathbf{L}f)_j = \int_0^\infty f(z)\,\mathrm{e}^{-t_j\,z}\,\mathrm{d}z\,, \quad j = 1,\ldots,m\,, \tag{16.10}$$

where $\Psi_m : \mathcal{C}(\mathbb{R}^+) \to \mathbb{R}^m$ denotes the point evaluations in t_j,

$$(\Psi_m g)_j = g(t_j)\,, \quad j = 1,\ldots,m\,. \tag{16.11}$$

Note that $\mathbf{L}f$ is in $\mathcal{C}^\infty(\mathbb{R}^+)$ for all $f \in L^2(\mathbb{R}^+)$ since

$$\exp(-t\,z) \in \mathcal{C}^\infty(\mathbb{R}^+ \times \mathbb{R}^+)\,.$$

Hence the *semi-discrete Laplace transform* $\mathbf{L}_m$ is well defined and the inequality

$$\|\mathbf{L}_m f\|_2^2 \leq \left(\sum_{j=1}^m \frac{1}{2\,t_j}\right) \|f\|_{L^2(\mathbb{R}^+)}^2$$

shows that $\mathbf{L}_m$ is a continuous, compact operator between the spaces $L^2(\mathbb{R}^+)$ and $\mathbb{R}^m$, whenever $t_j \neq 0$ for $j = 1,\ldots,m$. To get convergence results, we endow $\mathbb{R}^m$ with a weighted Euclidean scalar product

$$(x,y)_w := \sum_{j=1}^m w_j\,x_j\,y_j\,,$$

where the weights w_j are given as

$$w_j = \begin{cases} h_1/2\,, \text{if } j = 1\,, \\ (h_{j-1} + h_j)/2\,, \text{if } 1 < j < m\,, \\ h_{m-1}/2\,, \text{if } j = m \end{cases} \tag{16.12}$$

with $h_j = t_{j+1} - t_j$. If the scanning points t_j are equidistant, this means $t_j = a + (j-1)\,b/(m-1)$, we have $w_j = (b-a)/(m-1)$. Defining $\mathbf{L}_m$ as a mapping between $L^2(\mathbb{R}^+)$ and $\hat{\mathbb{R}}^m := (\mathbb{R}^m, (\cdot,\cdot)_w)$, the adjoint $\mathbf{L}_m^* : \hat{\mathbb{R}}^m \to L^2(\mathbb{R}^+)$ computes as

$$\mathbf{L}_m^* v(z) = \sum_{j=1}^m w_j\,v_j\,\mathrm{e}^{-t_j\,z}\,. \tag{16.13}$$

Thus, applying the trapezoidal sum corresponding to the nodes $\{t_j\}$ to $\mathbf{L}^* g$ gives $\mathbf{L}_m^*\,\Psi_m g$, when $g \in \mathcal{C}(\mathbb{R}^+) \cap L^2(\mathbb{R}^+)$.

The problem which we investigate, reads: for given data $g_m \in \hat{\mathbb{R}}^m$ we have to find $f \in L^2(\mathbb{R}^+)$ satisfying

$$\mathbf{L}_m f = g_m . \tag{16.14}$$

Because of the smoothing property of $\mathbf{L}$, it is obvious that (16.14) is a *severely ill-posed* problem in the sense, that any discretization of (16.14) leads to a matrix whose condition number grows exponentially in m.

We briefly formulate the concept of approximate inverse to solve (16.14). Let $e_\gamma \in L^2(\mathbb{R}^+ \times \mathbb{R}^+)$ be a mollifier, that means the mean value of $e_\gamma(\cdot, y)$ is equal to 1 for all $y \in \mathbb{R}^+$ and $\lim_{\gamma \to 0} \int f(z)\, e_\gamma(z, y)\, dz = f(y)$ a.e. Furthermore, suppose that $e_\gamma(\cdot, y)$ lies in the range $\mathsf{R}(\mathbf{L}_m^*)$[1] of $\mathbf{L}_m^*$. This implies the existence of a $v_\gamma(y)$ with

$$\mathbf{L}_m^* v_\gamma(y) = e_\gamma(\cdot, y) . \tag{16.15}$$

for fixed $y \in \mathbb{R}^+$. The moments $f_\gamma(y) := \langle f, e_\gamma(\cdot, y)\rangle_{L^2}$ are then computed by the semi-discrete approximate inverse $\widetilde{\mathbf{L}}_{m,\gamma} : \hat{\mathbb{R}}^m \to L^2(\mathbb{R}^+)$,

$$\widetilde{\mathbf{L}}_{m,\gamma} g_m(y) = (g_m, v_\gamma(y))_w .$$

In general, equation (16.15) is not solvable. The range $\mathsf{R}(\mathbf{L}_m^*)$ has finite dimension and is given by

$$\mathsf{R}(\mathbf{L}_m^*) = \text{span}\,\{\exp(-t_j \,\cdot) : j = 1, \ldots, m\} ,$$

see (16.13). All elements in $\mathsf{R}(\mathbf{L}_m^*)$ have exponential decay and (16.15) has a solution, only if $e_\gamma(\cdot, y)$ is a linear combination of functions of the type $\exp(-t_j \,\cdot)$. Moreover neither an inversion formula, nor a singular value decomposition is available for $\mathbf{L}_m$. Thus, we have to take a more careful choice of the mollifier than in the applications of Part II and III. We want to design a mollifier of exponential decay, which is adjusted to $\mathsf{R}(\mathbf{L}_m^*)$. To do so, we take an element ϕ_v^m from $\mathsf{R}(\mathbf{L}_m^*)$,

$$\phi_v^m(z) = \mathbf{L}_m^* v(z) = \sum_{j=1}^{m} w_j\, v_j\, \mathrm{e}^{-t_j\, z} \tag{16.16}$$

with $v \in \hat{\mathbb{R}}^m$ and w_j as in (16.12). Obviously ϕ_v^m in general does not fulfill the mollifier properties. So, we look for a bounded, linear mapping $\mathbf{A}_{y,\gamma}$, which acts on $L^2(\mathbb{R}^+)$ and forces ϕ_v^m to be a mollifier.

Lemma 16.1. *Let $\mathbf{A}_{y,\gamma} : L^2(\mathbb{R}^+) \to L^2(\mathbb{R}^+)$ be a bounded, linear mapping, $y > 0$, with*

$$\lim_{\gamma \to 0} \mathbf{A}_{y,\gamma}^* f(z) = f(y) \quad \textit{for almost every } z \in \mathbb{R}^+ , \tag{16.17}$$

[1] That means for fixed $y \in \mathbb{R}^+$ we postulate $e_\gamma(x, y) \in \mathsf{R}(\mathbf{L}_m^*)$ as a function of x.

and assume, that ϕ_v^m with (16.16) *fulfills the standardization*

$$\sum_{j=1}^{m} w_j\, v_j\, t_j^{-1} = 1\,. \tag{16.18}$$

Then

$$\lim_{\gamma\to 0}\langle \mathbf{A}_{y,\gamma}\phi_v^m, f\rangle_{L^2(\mathbb{R}^+)} = f(y) \quad a.e. \tag{16.19}$$

Remark 16.2. Standardization (16.18) is satisfied e.g., if $v_j = t_j\, w_j^{-1}/m$. Note that $w_j \neq 0$, since $t_j \neq t_k$ whenever $j \neq k$.

Proof. A quick calculation using (16.17) shows

$$\lim_{\gamma\to 0}\langle \mathbf{A}_{y,\gamma}\phi_v^m, f\rangle_{L^2(\mathbb{R}^+)} = \lim_{\gamma\to 0}\langle \phi_v^m, \mathbf{A}_{y,\gamma}^* f\rangle_{L^2(\mathbb{R}^+)} = f(y)\int_0^\infty \phi_v^m(z)\,\mathrm{d}z = f(y)\,,$$

since

$$\int_0^\infty \phi_v^m(z)\,\mathrm{d}z = \sum_{j=1}^{m} w_j\, v_j \int_0^\infty \mathrm{e}^{-t_j z}\,\mathrm{d}z = \sum_{j=1}^{m} w_j\, v_j\, t_j^{-1} = 1\,.$$

The last identity follows from (16.18). □

Thus, every mapping $\mathbf{A}_{y,\gamma}$ with (16.17) and (16.18) leads to a mollifier. Theorem 16.3 presents a special choice of $\mathbf{A}_{y,\gamma}$.

Theorem 16.3. *Assume that $\mathbf{A}_{y,\gamma}^* f(z) = f(\gamma\, y\, z + y)$ for γ, $y > 0$ and* (16.18) *holds. Then $e_\gamma^m(z,y) := \mathbf{A}_{y,\gamma}\phi_v^m(z)$ is a mollifier. Furthermore, defining for $y > 0$ a dilation D^y on $L^2(\mathbb{R}^+)$ by*

$$D^y f(z) = y^{-1} f(y^{-1}\, z)\,,$$

we get $e_\gamma^m(z,y) = D^y e_\gamma^m(z,1)$.

Proof. Obviously, (16.17) is valid, if we put $\mathbf{A}_{y,\gamma}^* f(z) = f(\gamma\, y\, z + y)$. Thus, because of Lemma 16.1, we only have to prove the normalization property. We first compute $\mathbf{A}_{y,\gamma}$ explicitly. With f, $g \in L^2(\mathbb{R}^+)$, we have

$$\begin{aligned}
\langle \mathbf{A}_{y,\gamma}^* f, g\rangle_{L^2(\mathbb{R}^+)} &= \int_0^\infty f(\gamma\, y\, z + y)\, g(z)\,\mathrm{d}z \\
&= \gamma^{-1} y^{-1} \int_0^\infty f(z + y)\, g(\gamma^{-1}\, y^{-1}\, z)\,\mathrm{d}z \\
&= \gamma^{-1} y^{-1} \int_y^\infty f(z)\, g(\gamma^{-1}\, y^{-1}\,(z - y))\,\mathrm{d}z \\
&= \langle f, \mathbf{A}_{y,\gamma} g\rangle_{L^2(\mathbb{R}^+)}\,,
\end{aligned}$$

and thus $\mathbf{A}_{y,\gamma} g(z) = \gamma^{-1} y^{-1} (1 - \chi_{[0,y)}(z))\, g(\gamma^{-1} y^{-1} z - \gamma^{-1})$, where $\chi_{[0,y)}$ means the characteristic function of the interval $[0, y)$. This yields

$$\int_0^\infty e_\gamma^m(z,y)\,\mathrm{d}z = \int_0^\infty \mathbf{A}_{y,\gamma} \phi_v^m(z)\,\mathrm{d}z$$

$$= (\gamma y)^{-1} \int_y^\infty \phi_v^m\Big((\gamma y)^{-1}(z-y)\Big)\,\mathrm{d}z = \int_0^\infty \phi_v^m(z)\,\mathrm{d}z = 1\,.$$

To show the dilation property, we use once more the representation of $\mathbf{A}_{y,\gamma}$ and get

$$\mathbf{A}_{y,\gamma} g(z) = D^y\{\gamma^{-1}(1-\chi_{[0,1)})\, g(\gamma^{-1}\cdot - \gamma^{-1})\}(z) = D^y\, \mathbf{A}_{1,\gamma} g(z)\,,$$

which completes the proof. □

Remark 16.4. Note that in Theorem 16.3 we designed a mollifier by forcing an element from $\mathsf{R}(\mathbf{L}_m^*)$ to be a mollifier. The resulting mollifier e_γ, however, is not an element from $\mathsf{R}(\mathbf{L}_m^*)$, but has the same exponential decay promising a good approximation by a function $\mathbf{L}_m^* v_\gamma(y)$.

The dilation property of the mollfier e_γ^m is associated with the dilation invariance of the Laplace transform $\mathbf{L}$. It is easy to check, that

$$\mathbf{L}\, D^y f = (D^y)^* \,\mathbf{L} f \quad \text{with} \quad (D^y)^* g(z) = g(y\,z)\,.$$

Figure 16.2 displays the mollifier e_γ^m from Theorem 16.3 for three different values of y.

To solve (16.15) numerically, we apply a collocation method. The possibility of using projection methods to compute reconstruction kernels was mentioned in Section 2.1. We choose collocation points z_k, $k = 1, \ldots, m$ and postulate (16.15) to hold in z_k,

$$\mathbf{L}_m^* v_\gamma(y)(z_k) = e_\gamma^m(z_k, y)\,, \quad k = 1, \ldots, m\,. \tag{16.20}$$

The system of linear equations (16.20) can be reformulated as

$$\mathbf{C}_m v_\gamma(y) = d_\gamma^m(y)\,, \tag{16.21}$$

where $\{\mathbf{C}_m\}_{k,j} = w_j \exp(-t_j\, z_k) \in \mathbb{R}^{m\times m}$ and $d_\gamma^m(y) \in \mathbb{R}^m$ with

$$d_\gamma^m(y)_k = e_\gamma^m(z_k, y) = \gamma^{-1} y^{-1} (1 - \chi_{[0,y)}(z_k))\, \phi_v(\gamma^{-1} y^{-1} z_k - \gamma^{-1})\,.$$

Theorem 16.5. *Setting*

$$z_k = \ell_1 + (k-1)\, h_m\,, \quad k = 1, \ldots, m\,, \qquad h_m = (\ell_2 - \ell_1)/(m-1) \tag{16.22}$$

uniformly in some interval $[\ell_1, \ell_2]$ *leads to an invertible matrix* $\mathbf{C}_m$.

Proof. To show the non-singularity of $\mathbf{C}_m$, it suffices to prove the injectivity of the transpose $\mathbf{C}_m^\top$. Let $v \in \mathbb{R}^m$ be an element from the kernel of $\mathbf{C}_m^\top$, that means $\mathbf{C}_m^\top v = 0$. Then

$$\sum_{k=1}^{m} w_j\,\mathrm{e}^{-t_j\,z_k}\,v_k = w_j\,\mathrm{e}^{-t_j\,\ell_1}\sum_{k=1}^{m}\left(\mathrm{e}^{-t_j\,h_m}\right)^{k-1} v_k = 0$$

yielding

$$\sum_{k=1}^{m}\left(\mathrm{e}^{-t_j\,h_m}\right)^{k-1} v_k = 0\,, \quad \text{for } j = 1,\ldots,m\,. \tag{16.23}$$

System (16.23) is a Vandermonde system and hence regular, which gives $v = 0$. □

For the remainder of this section we suppose that *the collocation points are equidistant* according to Theorem 16.5, which implies that $\mathbf{C}_m$ has an inverse $\mathbf{C}_m^{-1}$. Since system (16.21) is the discretization of a severely ill-posed problem, we expect $\mathbf{C}_m$ to be extremely ill-conditioned. So, to get a stable solution, regularization is necessary, even though the right hand side of (16.21) contains no noise. Thus, instead of (16.21) we consider the system

$$(\mathbf{C}_m + \rho\,I_m)\,v_\gamma^\rho(y) = d_\gamma^m(y)\,, \tag{16.24}$$

which arises from (16.21) applying Lavrentiev's method. Here, I_m denotes the identity matrix in $\mathbb{R}^m$. Since in general $\mathbf{C}_m$ is indefinite, we have to find a criterion for ρ resulting in a stably invertible matrix

$$\mathbf{C}_m^\rho := \mathbf{C}_m + \rho\,I_m\,.$$

Theorem 16.6. *Let $\|\cdot\|$ be a matrix norm and $\rho \in \mathbb{R}$ with $|\rho| < \|\mathbf{C}_m^{-1}\|^{-1}/2$. Then $\mathbf{C}_m^\rho$ is invertible and*

$$\|(\mathbf{C}_m^\rho)^{-1}\| < |\rho|^{-1}\,. \tag{16.25}$$

Proof. Because of $|\rho|\,\|\mathbf{C}_m^{-1}\| < 1/2$, we have from Neumann's series

$$\|(I_m + \rho\,\mathbf{C}_m^{-1})^{-1}\| = \left\|\sum_{k=0}^{\infty}(-\rho\,\mathbf{C}_m^{-1})^k\right\| = \frac{1}{1 - |\rho|\,\|\mathbf{C}_m^{-1}\|} \le 2\,.$$

Assertion (16.25) follows from

$$\|(\mathbf{C}_m^\rho)^{-1}\| = \|(I_m + \rho\,\mathbf{C}_m^{-1})^{-1}\,\mathbf{C}_m^{-1}\| \le 2\,\|\mathbf{C}_m^{-1}\| < |\rho|^{-1}\,.$$

□

Condition

$$|\rho| < \|\mathbf{C}_m^{-1}\|^{-1}/2 \tag{16.26}$$

guarantees the solvability of (16.24), but means a very strong restriction on ρ, because commonly $\|\mathbf{C}_m^{-1}\|$ is very large. Since condition (16.26) is not necessary, but only sufficient, and $\det(\mathbf{C}_m^\rho) = 0$ for finitely many ρ only, we expect that $\mathbf{C}_m^\rho$ has an inverse even for some $|\rho| \geq \|\mathbf{C}_m^{-1}\|^{-1}/2$. Figure 16.3 shows the three reconstruction kernels $v_\gamma^\rho(y)$ associated with the mollifiers $e_\gamma^m(\cdot, y)$ in Figure 16.2. The kernels were computed as solutions from (16.24) with $\rho = 10^{-15}$. Note that the reconstruction kernels have the same dilation property as the mollifiers e_γ^m mentioned in Theorem 16.3. The mollifier $e_\gamma^m(\cdot, y)$ for $y = 0.5$ is plotted in Figure 16.4 together with its approximation $\mathbf{L}_m^* v_\gamma^\rho(y)$ in $\mathsf{R}(\mathbf{L}_m^*)$. We see that $\mathbf{L}_m^* v_\gamma^\rho(y)$ also has its global maximum at $z = 0.5$ and that the approximation becomes worse as $z \to 0$. This comes from the exponential increase of $\exp(-t_j\, z)$ for $z \to -\infty$.

Our inversion scheme for $\mathbf{L}_m$ now reads as follows: for given data $g_m \in \mathbb{R}^m$ compute $\widetilde{\mathbf{L}}_{m,\gamma}^\rho g_m$

$$\widetilde{\mathbf{L}}_{m,\gamma}^\rho g_m(y) = (g_m, v_\gamma^\rho(y))_w\,. \tag{16.27}$$

In [111, Section 4], the author presented an error estimate for (16.27). We will state the most important results. Let f having compact support in some interval $[\ell_1, \ell_2]$. Without loss of generality we may assume that $0 < \ell_1 < \ell_2$ what always can be achieved using

$$\mathbf{L}f(t) = \mathrm{e}^{r\,t}\, \mathbf{L}\{f(\cdot - r)\}(t)\,.$$

As was done to show convergence in case of the Doppler transform, see Corollary 8.5 in Chapter 8, we have to couple the data scanning in t and the regularization parameters γ and ρ. With

$$\delta_m := \max\{t_j - t_{j-1} : j = 2, \ldots, m\}\,,$$

we denote the maximal discretization step size. First, we estimate

$$|f_\gamma(y) - \widetilde{\mathbf{L}}_{m,\gamma}^\rho\, \mathbf{L}_m f(y)|\,.$$

As a result of our investigations we will see that the error essentially is a sum of three parts: A regularization error, a discretization error and the error coming from applying Lavrentiev's method (16.21). The proof of Theorem 16.7 is omitted; it can be found in [111].

Theorem 16.7. *Let $f \in H^\alpha(\ell_1, \ell_2)$, $1/2 < \alpha \leq 1$ with supp $f \subset [\ell_1, \ell_2]$, $y \in (\ell_1, \ell_2)$ be a reconstruction point, e_γ^m be the mollifier designed in Theorem* 16.3 *and $\gamma > 0$ be a fixed regularization parameter. Furthermore, we assume $\{\rho_m\}$ to be a sequence with* $\lim_{m\to\infty} \rho_m = 0$ *and* $0 < \rho_m < \|(\mathbf{C}_m)^{-1}\|^{-1}/(m+1)$. *Then*

$$\begin{aligned} |f_\gamma(y) - \langle \mathbf{L}_m f, v_\gamma^{\rho_m}(y)\rangle_{\mathbb{R}^m}| \leq \Big(&C_v\, \gamma^{-1}\, y^{-1}\, m^{-1} \\ &+ h_m^\alpha\, \{C_{1,\alpha}\, \gamma^{-\frac{\alpha+1}{2}}\, y^{-\frac{\alpha+1}{2}} + C_{2,\alpha}\, \gamma^{-1}\, y^{-1}\, \delta_m\, \rho_m^{-1}\}\Big) \|f\|_{H^\alpha(\ell_1,\ell_2)} \end{aligned} \tag{16.28}$$

for $m \to \infty$. *If* δ_m *satisfies* $\delta_m \le c\, h_m^\lambda\, \rho_m$ *with* $\lambda < \alpha$, $c > 0$ *for* $m \to \infty$, *we get for a fixed* $\gamma > 0$ *the convergence result*

$$\lim_{m\to\infty} |f_\gamma(y) - \langle \mathbf{L}_m f, v_\gamma^{\rho_m}(y)\rangle_{\hat{\mathbb{R}}^m}| = 0 \,. \tag{16.29}$$

Since we know from Theorem 16.3 that

$$|f(y) - f_\gamma(y)| \le \kappa_\gamma(y) \tag{16.30}$$

for a function κ_γ with $\lim_{\gamma\to 0} \kappa_\gamma(y) = 0$ for almost every y, we get the following convergence estimate from Theorem 16.7.

Corollary 16.8. *Adopt all assumptions from Theorem* 16.7 *and assume the sampling condition* $\delta_m \le c\, h_m^\lambda\, \rho_m$ *with* $\lambda < \alpha$ *for* $m \to \infty$. *Then*

$$\lim_{m\to\infty} |f(y) - f_\gamma^{\rho_m}(y)| \le \kappa_\gamma(y)$$

for almost every $y \in (\ell_1, \ell_2)$.

Proof. The proof follows from Theorem 16.7, (16.30) and an application of the triangle inequality

$$|f(y) - f_\gamma^{\rho_m}(y)| \le |f(y) - f_\gamma(y)| + |f_\gamma(y) - f_\gamma^{\rho_m}(y)| \,.$$

Note that by (16.27) $f_\gamma^{\rho_m}(y) = (\mathbf{L}_m f, v_\gamma^{\rho_m}(y))_w = \langle f, \mathbf{L}_m^* v_\gamma^{\rho_m}(y)\rangle_{L^2(\ell_1,\ell_2)}$ holds true. □

Remark 16.9. Condition $\delta_m \le c\, h_m^\lambda\, \rho_m$, $\lambda < \alpha$ implies that the maximal scanning step size converges to zero as m grows to infinity. Convergence estimate (16.29) is valid only if there exists a sampling δ_m fulfilling that condition.

So far, we have proved a limit estimate for a *fixed* regularization parameter $\gamma > 0$. Now we present conditions on a sequence of parameters $\{\gamma_m\}$ leading to point-wise convergence of the approximate inverse to the exact solution f.

Theorem 16.10. *Adopt all assumptions made in Theorem* 16.7. *Let* $\{\gamma_m\}$ *be a sequence of positive numbers with* $\lim_{m\to\infty} \gamma_m = 0$ *such that*

$$\lim_{m\to\infty} \gamma_m^{-1}\, m^{-1} = \lim_{m\to\infty} h_m^\alpha\, \gamma_m^{-\frac{\alpha+1}{2}} = \lim_{m\to\infty} h_m^\alpha\, \gamma_m^{-1}\, \delta_m\, \rho_m^{-1} = 0 \,. \tag{16.31}$$

Then we have for almost every $y \in (\ell_1, \ell_2)$ *the convergence*

$$\lim_{m\to\infty} |f(y) - f_{\gamma_m}^{\rho_m}(y)| = 0 \,. \tag{16.32}$$

A sequence $\{\gamma_m\}$ *satisfying* (16.31) *is given by*

$$\gamma_m = \frac{1}{(\ln m)^\tau} \,, \qquad \tau > 0$$

where δ_m *fulfills* $\delta_m \le c\, m^\lambda\, \rho_m$ *for a constant* $c > 0$ *and* $\lambda < \alpha$.

Proof. Condition (16.31) guarantees the convergence of the right-hand side of (16.28) to 0 for $m \to \infty$. Furthermore, $\lim_{m\to\infty} \kappa_{\gamma_m}(y) = 0$, since γ_m converges to 0. Thus, convergence (16.32) becomes clear from Corollary 16.8. Defining $\gamma_m = 1/(\ln m)^\tau$ for $\tau > 0$ and constraining δ_m by $\delta_m \le c\, m^\lambda \rho_m$ for a constant $c > 0$ and $\lambda < \alpha$ obviously satisfies the conditions (16.31). □

Remark 16.11. In view of Theorem 16.7 and Corollary 16.8, the reconstruction error consists of several parts: a regularization error depending only on γ, discretization errors depending on h_m and δ_m and a further regularization error which comes from applying Lavrentiev's method to (16.21). The discretization errors are the crucial ones, since they vanish for $m \to \infty$ only if we have a sampling condition as stated in Theorem 16.7.

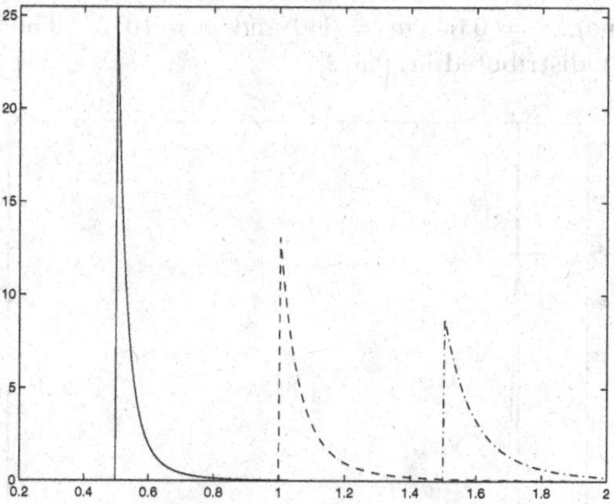

Fig. 16.2. Plot of $e_\gamma^m(\cdot, y) = \mathbf{A}_{y,\gamma}\phi_v^m$ from Theorem 16.3 for $\gamma = 0.04$, $y = 0.5$ (solid line), $y = 1.0$ (dashed line) and $y = 1.5$ (dashed-dotted line). The vector v is chosen according to Remark 16.2, where the scanning points t_j are equally distributed in $[0.3, 2]$, $m = 300$.

16.3 A solution scheme for the X-ray diffractometry problem

The solution scheme for X-ray diffractometry published in [114] consists of two steps: The computation of $\breve{\sigma}_{ij}$ from (16.5) (with σ_{ij} replaced by $\breve{\sigma}_{ij}$) and an inversion of the Laplace transform to get σ_{ij}. Let measured peak shifts

$$\varepsilon_{\varphi_l \psi_k} = -\cot\theta_0\,(\theta_{\varphi_l \psi_k} - \theta_0) \tag{16.33}$$

be given for a finite number of tilt angles ψ_k and rotation angles φ_l. We use a two-step method to regain the stress tensor from (16.33).

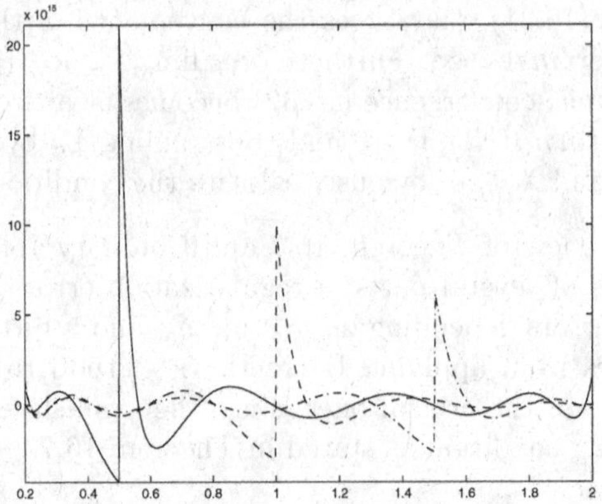

Fig. 16.3. Plot of $v_\gamma^\rho(y)$ for $y = 0.5$ (solid line), $y = 1.0$ (dashed line), $y = 1.5$ (dashed-dotted line), $\gamma = 0.04$, $m = 300$ and $\rho = 10^{-15}$. The collocation points $z_k = t_k$ are equally distributed in $[0.3, 2]$.

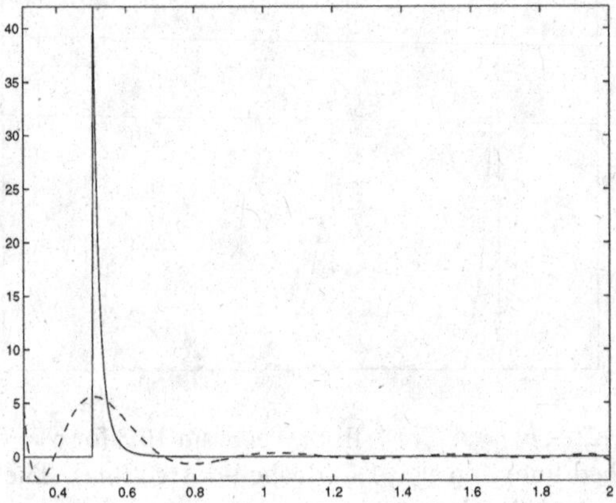

Fig. 16.4. Mollifier $e_\gamma^m(\cdot, y)$ for $y = 0.5$, $\gamma = 0.04$ (solid curve) and its approximation $\mathbf{L}_m^* v_\gamma^\rho(y)$ in $\mathsf{R}(\mathbf{L}_m^*)$ (dashed curve). Here, $v_\gamma^\rho(y)$ solves (16.24) with $\rho = 10^{-15}$, $m = 300$.

'Figures 16.2 – 16.4 have been reproduced from SCHUSTER [111] with kind permission of de Gruyter GmbH & Co. KG.'

Step 1: Solve (16.5) by the method of least squares and get $\check{\sigma}_{ij}$.

Step 2: Calculate

$$\sigma_{ij}(z) = \left(\check{\sigma}_{ij}(\tau_{\psi_k}), \tau_{\psi_k}\, v_\gamma^\rho(z; \varphi_l, \psi_k)\right)_w$$

according to (16.27). Here ψ_k, φ_l are the angles for which measured data are acquired and the reconstruction kernel v_γ^ρ is a solution of (16.24). The additional factor τ_{ψ_k} comes from the reciprocal Laplace transform (16.6).

The method was used to recover the stress profile of a ground corundum (Al_2O_3) specimen. The measurement used 44 tilt angles ψ_k and the two rotation angles $\varphi_1 = 0°$, $\varphi = 90°$ which gives a total number of 88 data values. The maximum tilt angle was $\psi_{\max} = 87°$ and the measure process was carried out with synchroton radiation at HASYLAB (DESY), Hamburg (Germany). The wave length of the applied X-ray beams was $\lambda = 179$ pm and the Bragg angle of the unstressed probe $\theta_0 = 37.1°$ was determined by averaging the measured Bragg angles $\theta_{\varphi_l \psi_k}$. Figure 16.5 shows the measured reflection angles at the different tilt angles and the recovered stress tensor components σ_{11} and σ_{22}. The regularization parameters were $\gamma = 0.035$, $\rho = 0.1$. Hence, the method of approximate inverse led to a stable solver in the bad situation of a severely ill-posed problem with only a few number of (noisy) data.

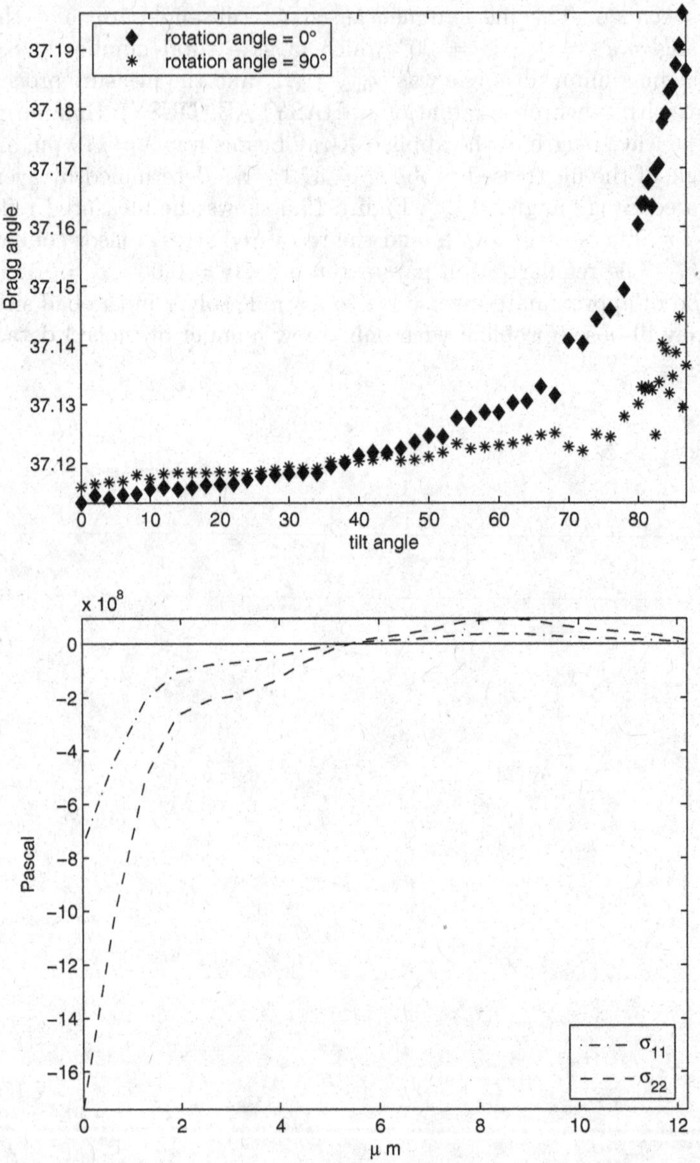

Fig. 16.5. Upper picture: Measured Bragg angles at different tilt angles varying from 0° to 87° for the two rotations $\varphi_1 = 0°$ and $\varphi_2 = 90°$. Lower picture: Calculated stress profile of the components σ_{11} (dashed-dotted line) and σ_{22} (dashed line) depending on depth. The method of approximate inverse was applied using the parameters $\gamma = 0.035$, $\rho = 0.1$.

17

A filtered backprojection algorithm for thermoacoustic computerized tomography (TCT)

Thermoacoustic computerized tomography (TCT) is a novel imaging technique for non-destructive testing and medical imaging, which combines both the advantages of purely optical imaging (high contrast) and ultrasound imaging (high resolution). TCT uses electromagnetic energy as input and the induced thermoacoustic pressure field as measurement output. The determination of the unknown energy deposition function is equivalent to the reconstruction of a function from its integrals over spheres. In TCT the center sets are spheres whereas the SONAR and SAR problems usually use hyperplanes as center sets. This leads to a spherical mean operator with different mathematical properties. We apply the method of approximate inverse to derive a reconstruction algorithm of filtered backprojection type.

17.1 Thermoacoustic computerized tomography (TCT)

The aim of TCT is the reconstruction of spatial inhomogeneities in human tissue or some specimen. To this end, one illuminates the sample by a pulsed electromagnetic energy resulting in a non-uniform energy deposition within the sample followed by a thermoelastic expansion which causes an acoustic pressure wave, see GUSEV AND KARABUTOV [35], and LIU [93]. Measuring the induced thermoacoustic pressure wave with the help of acoustic detectors which surround the object, TCT tries to recover the energy deposition function. A scheme of the scanning system is sketched in Figure 17.1. The energy deposition itself depends on the absorptivity which in return offers clues to the searched inhomogeneities. HALTMEIER ET AL. [37] developed a measurement setup and reconstruction method which is based on the Radon transform using large planar receivers. In [92] PATCH delivers consistency conditions which are useful when part of the data is unknown. The articles of KRUGER ET AL. [60] and XU ET AL. [133, 132, 134] also contain basics of TCT. An inversion formula for TCT using a series expansion was presented by NORTON AND LINZER [89]. Inversion formulas for the spherical mean operator in odd dimensions as well as uniqueness results are found in FINCH, PATCH, AND RAKESH [31]. We follow the outlines of HALTMEIER ET AL. [38] and present an inversion scheme based on the method of approximate inverse.

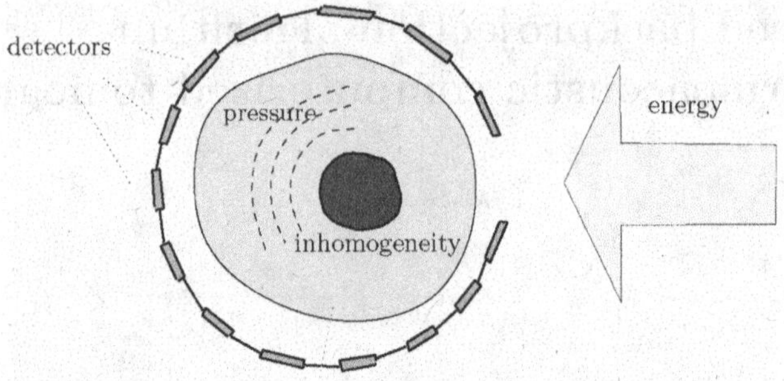

Fig. 17.1. Thermoacoustic scanning system. The examined object is illuminated by a short electromagnetic pulse. The *thermoelastic effect* (c.f. Figure 17.2) causes an evolving pressure wave that is measured with several acoustic detectors enclosing the imaged object.

'The picture was reprinted from M. HALTMEIER, T. SCHUSTER, AND O. SCHERZER, *Filtered backprojection for thermoacoustic computed tomography in spherical geometry*, Math. Meth. Appl. Sci., 28 (2005), pp. 1919–1937. Copyright ©2005 John Wiley & Sons Limited. Reproduced with permission.'

The specimen under consideration is illuminated by a pulsed electromagnetic signal $j(t)$. The temporal shape of that signal is assumed to have small support $[0, \tau]$, where $\tau \ll 1$. In fact, that pulse lasts a few picoseconds only. The signal j is furthermore supposed to be non-negative, smooth (at least $j \in C^1(\mathbb{R})$) and to satisfy

$$\int_{\mathbb{R}} j(t)\,\mathrm{d}t = 1\,.$$

That means, j has the shape of a delta peak. The absorbed energy per unit volume and unit time $r(x,t)$ is given as

$$r(x,t) = I_{\mathrm{em}}(x)\,j(t)\,\psi(x)\,,$$

where I_{em} denotes the radiation intensity and $\psi(x)$ is the absorption density inside the fluid which surrounds the object under investigation. The heating caused by the absorbed energy leads to an expansion of the volume. The rate of change of the volume depends on the thermal expansion coefficient β, the adiabatic speed of sound v_s, the specific heat capacity c_p of the material and is related with the rate of change of pressure $p(x,t)$ and the absorbed energy $r(x,t)$ as

$$\frac{\partial \varrho}{\partial t} = \frac{1}{v_s^2}\frac{\partial p}{\partial t} - \frac{\beta}{c_p}\,r\,. \tag{17.1}$$

Equation (17.1) is called the *expansion equation*. The *thermoelastic effect* is described by this very equation: the absorbed energy causes a thermal expansion $\partial_t \varrho$ and a pressure field p. Figure 17.2 emphasizes that issue.

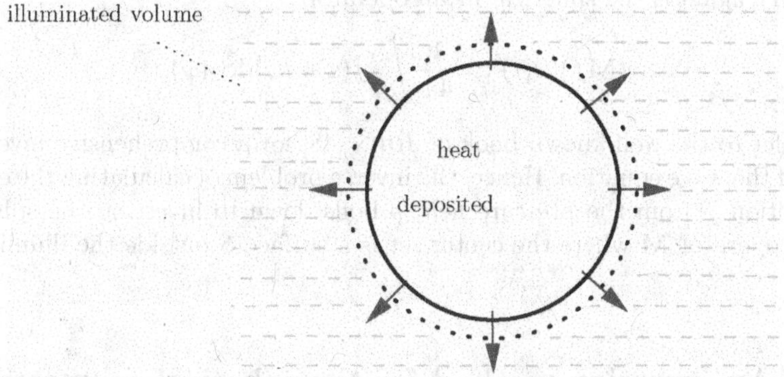

Fig. 17.2. Thermoelastic effect. The absorbed electromagnetic energy within the illuminated part of the fluid causes thermal expansion and a subsequent pressure field. The dependance between the thermoelastic expansion and the pressure on the received electromagnetic energy is given by the expansion equation (17.1).
'The picture was reprinted from M. HALTMEIER, T. SCHUSTER, AND O. SCHERZER, *Filtered backprojection for thermoacoustic computed tomography in spherical geometry*, Math. Meth. Appl. Sci., 28 (2005), pp. 1919–1937. Copyright ©2005 John Wiley & Sons Limited. Reproduced with permission.'

Together with some fundamental equations of fluid dynamics like the linearized continuity equation and Euler equation, see e.g. CHORIN, MARSDEN [16], LANDAU, LIFSCHITZ [61], equation (17.1) results in

$$\Big(\frac{\partial^2}{\partial t^2} - \Delta\Big)p = f(x)\, j'(t)\,, \tag{17.2}$$

where

$$f(x) .:= \frac{\beta\, v_s}{c_p}\, I_{\mathrm{em}}(x)\, \psi(x)$$

is the *energy deposition* function. A very detailed proof of relation (17.2) is found in [38]. The inverse problem of TCT consists of recovering the energy deposition function $f(x)$ from the measured pressure field $p(x,t)$ at the detectors.

Endowed with appropriate initial conditions, equation (17.2) has a unique solution. We use

$$p(x,0) = 0\,, \qquad \frac{\partial p}{\partial t}(x,0) = 0\,, \tag{17.3}$$

taking into account that there is no acoustic pressure before the experiment starts. The unique solution of (17.2), (17.3) is given by

$$p = j'(t) *_t (t\,\mathbf{M}f)\,, \tag{17.4}$$

where $*_t$ means the Laplace-convolution with respect to t

$$(g_1 *_t g_2)(x,t) := \int_0^t g_1(x, t-s)\, g_2(x,s)\, \mathrm{d}s$$

and $\mathbf{M}f$ again is the spherical mean operator

$$(\mathbf{M}f)(x,t) := \frac{1}{4\pi} \int_{S^2} f(x + t\omega) \mathrm{d}S_2(\omega)\,. \tag{17.5}$$

We refer to the well-known book of JOHN [48] for a comprehensive investigation of the wave equation. Hence, the inverse problem of calculating the energy deposition f from the pressure field p boils down to inverting the spherical mean operator $\mathbf{M}$ where the center set is a surface S outside the illuminated fluid.

17.2 An inversion method for the spherical geometry

In order to apply the method of approximate inverse to (17.5), we first prove some important results of the spherical mean operator as well as of some related operators. Note that the spherical mean operator in case of TCT differs from that of SONAR and SAR problems, because the center sets are different. In TCT we have the (comfortable) situation of a linear and bounded operator between Hilbert spaces and we need not to consider distribution spaces.

We explicitly deal with the case where the energy deposition function f is supported in a closed ball $B_\rho := \overline{B_\rho(0)}$ with center 0 and radius ρ and in which the thermoacoustic pressure field is measured on $S_\rho := \partial B_\rho$. Let

$$X_\rho := L^2(B_\rho) = \{f \in L^2(\mathbb{R}^3) : f(x) = 0 \text{ for a.e. } x \in \mathbb{R}^3 \setminus B_\rho\}$$

be the Hilbert space of all square integrable functions supported in B_ρ with inner product

$$\langle f_1, f_2 \rangle_\rho := \int_{B_\rho} f_1(x)\, f_2(x)\, \mathrm{d}x$$

and norm $\|\cdot\|_\rho$. For all $T \geq 0$, let

$$Y_{\rho,T} := L^2(S_\rho \times [0,T])$$

denote the Hilbert space of all square integrable functions $f : S_\rho \times [0,\infty) \to \mathbb{R}$ supported in $S_\rho \times [0,T]$ with inner product

$$\langle g_1, g_2 \rangle_{\rho,T} := \int_{S_\rho} \int_0^T g_1(\sigma,t)\, g_2(\sigma,t)\, \mathrm{d}t\, \mathrm{d}S_2^\rho(\sigma)$$

and associated norm $\|\cdot\|_{\rho,T}$. Here $\mathrm{d}S_2^\rho$ denotes the surface measure on S_ρ. We furthermore write D_t for the operator that maps $\varphi \in \mathcal{C}^1(S_\rho \times [0,\infty))$ onto its derivative with respect to the second variable

$$D_t\varphi(y,t) = \varphi_t(y,t)\,, \qquad y \in S_\rho\,, \quad t \in [0,\infty)\,.$$

Finally, we introduce operators

$$\begin{aligned} \mathbf{N} : \mathcal{C}_0(B_\rho) \subset X_\rho \to Y_{\rho,2\rho}\,, &\qquad (\mathbf{N}f)(\sigma,t) := t\mathbf{M}f(\sigma,t)\,, \\ \mathbf{P} : \mathcal{C}_0(B_\rho) \subset X_\rho \to Y_{\rho,2\rho+\tau}\,, &\qquad \mathbf{P}f = D_t j *_t \mathbf{N}f\,. \end{aligned}$$

From (17.4) we see that the operator $\mathbf{P}$ maps an unknown energy deposition function f onto the thermoacoustic pressure field restricted to the recording surface S_ρ. Particularily, if f is a $\mathcal{C}^1$-function and the pulse duration τ tends to 0, then $\mathbf{P}f$ tends to $D_t\mathbf{N}f$. Furthermore $\mathbf{N}$, $\mathbf{P}$ are linear bounded operators between L^2-spaces.

Lemma 17.1. *Let $f \in \mathcal{C}_0(B_\rho)$. Then $\|\mathbf{N}f\|^2_{\rho,2\rho} \le \rho^2\,\|f\|^2_\rho$ and*

$$\|\mathbf{P}f\|^2_{\rho,2\rho+\tau} \le (2\rho+\tau)\,\tau\,\rho^2\|D_t j\|^2_\infty\,\|f\|^2_\rho\,. \tag{17.6}$$

Here, $\|D_t j\|_\infty := \sup\{|D_t j(t)| : 0 \le t \le \tau\}$ denotes the supremum norm of $D_t j$.

Proof. Let $\sigma \in S_\rho$. Since $\operatorname{supp} f \subset B_\rho$, an application of the Cauchy-Schwartz inequality shows

$$\begin{aligned} \|\mathbf{N}f(\sigma,\cdot)\|^2_{L^2[0,2\rho]} &= \int_0^{2\rho} \left(\frac{t}{4\pi}\int_{S^2} f(\sigma + t\,\omega)\,\mathrm{d}S_2(\omega)\right)^2 \mathrm{d}t \\ &\le \frac{1}{4\pi}\int_0^{2\rho}\int_{S^2} f(\sigma+t\,\omega)^2\,\mathrm{d}S_2(\omega)\,t^2\,\mathrm{d}t = \frac{1}{4\pi}\,\|f\|^2_\rho\,, \end{aligned}$$

whence

$$\|\mathbf{N}f\|^2_{\rho,2\rho} = \int_{S_\rho} \|\mathbf{N}f(\sigma,\cdot)\|^2_{L^2[0,2\rho]}\,\mathrm{d}S_2^\rho(\sigma) \le \rho^2\,\|f\|^2_\rho\,. \tag{17.7}$$

Next we verify (17.6). Assume $\sigma \in S_\rho$ and $t \in [0, 2\rho+\tau]$. Again, we use the Cauchy-Schwartz inequality to obtain

$$\begin{aligned} |D_t j *_t \mathbf{N}f(\sigma,t)|^2 &= \left[\int_0^t \mathbf{N}f(\sigma,t-s)(D_t j)(s)\,\mathrm{d}s\right]^2 \\ &\le \|\mathbf{N}f(\sigma,\cdot)\|^2_{L^2[0,2\rho]}\int_0^\tau (D_t j)(s)^2\,\mathrm{d}s \le \tau\,\|D_t j\|^2_\infty\,\|\mathbf{N}f(\sigma,\cdot)\|^2_{L^2[0,2\rho]}\,. \end{aligned}$$

From the last inequality and (17.7) we conclude that

$$\begin{aligned}\|\mathbf{P}f\|^2_{\rho,2\rho+\tau} &= \int_{S_\rho} \int_0^{2\rho+\tau} |D_t j *_t \mathbf{N}f(\sigma,t)|^2 \,\mathrm{d}t \,\mathrm{d}S_2(\sigma) \\ &\le (2\rho+\tau)\,\tau\, \|D_t j\|^2_\infty \int_{S_\rho} \|\mathbf{N}f(\sigma,\cdot)\|^2_{L^2[0,\tau]} \\ &\le (2\rho+\tau)\,\tau\,\|D_t j\|^2_\infty\, \rho^2\, \|f\|^2_\rho\,,\end{aligned}$$

proving (17.6).

As a consequence of Lemma 17.1, the operators $\mathbf{N}$ and $\mathbf{P}$ extend in a unique way to bounded linear operators $\mathbf{N} : X_\rho \to Y_{\rho,2\rho}$ and $\mathbf{P} : X_\rho \to Y_{\rho,2\rho+\tau}$, respectively. In particular, they have bounded adjoints. The proof of Lemma 17.2 is found in [38].

Lemma 17.2. *Let $g \in \mathcal{C}_0(S_\rho \times [0,2\rho])$ and $p \in \mathcal{C}_0(S_\rho \times [0,2\rho+\tau])$. Then*

$$(\mathbf{N}^* g)(x) = \frac{1}{4\pi}\int_{S_\rho} \frac{g(x,\|x-\sigma\|)}{\|x-\sigma\|}\,\mathrm{d}S_2^\rho(\sigma)\,, \qquad x \in B_\rho\,, \tag{17.8a}$$

$$(\mathbf{P}^* p)(x) = -\frac{1}{4\pi}\int_{S_\rho} \frac{(D_t\bar{j}\star p)(x,\|x-\sigma\|)}{\|x-\sigma\|}\,\mathrm{d}S_2^\rho(\sigma)\,, \quad x \in B_\rho\,. \tag{17.8b}$$

Here $\bar{j}$ is defined by $\bar{j}(s) := j(-s)$ and

$$(D_t\bar{j}\star p)(x,s) := \int_s^{s+\tau} D_t\bar{j}(s-t)p(x,t)\,dt\,.$$

Remark 17.3. The duration τ of the electromagnetic pulse is typically in the range of picoseconds. Hence, the temporal part of the electromagnetic pulse can be approximated by the delta distribution. Therefore we can regard $D_t\mathbf{N}f$ as measurement data instead of $\mathbf{P}f$. From Lemma 17.2, it follows that an appropriate approximation to the adjoint $\mathbf{P}^*$ of $\mathbf{P}$ is given by $-\mathbf{N}^* D_t$.

Finch et al. [31] proved the injectivity of the operator $\mathbf{N}$ on the space of smooth (i.e. $\mathcal{C}^\infty$-) functions which are supported in B_ρ and stated several inversion formulas for it.

Theorem 17.4. [31, Theorem 3] *Let $f \in \mathcal{C}_0^\infty(B_\rho)$. Then*

$$f = -\frac{2}{\rho}\mathbf{N}^* D_t\, t\, D_t\, \mathbf{N}f\,. \tag{17.9}$$

Formula (17.9) even is valid, if we have weaker smoothness assumptions.

Corollary 17.5. *Let $f \in \mathcal{C}_0^1(B_\rho)$ and assume $D_t\mathbf{N}f \in \mathcal{C}_0^1(S_\rho \times [0,2\rho])$. Then (17.9) holds true.*

Proof. Let $\varphi \in \mathcal{C}_0^\infty(B_\rho)$. Equation (17.9) holds true for φ and thus

$$-\rho/2\,\langle f,\varphi\rangle_\rho = \langle f, \mathbf{N}^*\, D_t\, t\, D_t\, \mathbf{N}\varphi\rangle_\rho\,.$$

Since f and hence $\mathbf{N}^* f$ is a $\mathcal{C}^1$ function the right-hand side of this equation equals to $-\langle t\, D_t\, \mathbf{N} f, D_t\, \mathbf{N}\varphi\rangle_{\rho,2\rho}$. Together with the assumption $D_t\, \mathbf{N} f \in \mathcal{C}_0^1(S_\rho \times [0,2\rho])$, this yields

$$-\frac{\rho}{2}\,\langle f,\varphi\rangle_\rho = \langle \mathbf{N}^*\, D_t\, t\, D_t\, \mathbf{N} f, \varphi\rangle_\rho\,.$$

Since this equation is valid for all $\varphi \in \mathcal{C}_0^\infty(B_\rho)$, we may conclude that $-\rho/2 f = \mathbf{N}^*\, D_t\, t\, D_t\, \mathbf{N} f$.

After these preliminaries, we proceed to apply the method of approximate inverse to the problem of TCT, i.e. the solution of

$$\mathbf{P} f = p\,.$$

Hence, instead of solving $\mathbf{P} f = p$ we search for smoothed approximations $(f_{\gamma,\nu})_{\gamma>0}$ with

$$f_{\gamma,\nu}(y) := \langle f, e_{\gamma,\nu}(\cdot,y)\rangle_\rho\,, \qquad y \in \mathbb{R}^3\,,$$

where we again consider radially symmetric mollifiers $e_{\gamma,\nu}(\cdot,y)$, $\gamma > 0$, of the form (2.3)

$$e_{\gamma,\nu}(x,y) = \frac{1}{I_\nu \gamma^3}\, R_\nu\!\left(\frac{\|y-x\|^2}{\gamma^2}\right), \qquad x,y \in \mathbb{R}^3\,. \tag{17.10}$$

Here, $\nu > 0$ is a real number, $I_\nu := \pi^{3/2}\,\Gamma(\nu+1)/\Gamma(\nu+5/2)$ is a scaling factor and R_ν denotes a function on $[0,\infty)$, defined by

$$R_\nu(s) := \begin{cases} (1-s)^\nu\,, & \text{if } 0 \le s \le 1\,, \\ 0\,, & \text{if } s \ge 1\,. \end{cases} \tag{17.11}$$

A plot of the radial part of $e_{\gamma,\nu}(\cdot,y)$ can be seen in the picture to the left in Figure 17.3.

Remark 17.6. We have chosen a shift invariant mollifier $e_{\gamma,\nu}$ (17.10) though the operators $\mathbf{P}$ and $\mathbf{N}$ are not translation invariant. The reason is simple: A mollifier generated by translations is associated to a reconstruction kernel which is also shift invariant, see representation (17.18).

Since supp $R_\nu = [0,1]$, the mollifier $e_{\gamma,\nu}(\cdot,y)$ has support $\overline{B_\gamma(y)}$ and the chain rule guarantees that $e_{\gamma,\nu}(\cdot,y) \in \mathcal{C}^k(\mathbb{R}^3)$ for all integer numbers $k < \nu$. Furthermore, the scaling factor I_ν has been chosen such that

$$\int_{B_\gamma(y)} e_{\gamma,\nu}(x,y)\,dx = 1\,, \qquad y \in \mathbb{R}^3\,, \tag{17.12}$$

which can be shown by a quick calculation. Hence, $e_{\gamma,\nu}$ as in (17.10) is a mollifier according to definition 2.1.

From Corollary 17.5, we deduce that if $e_{\gamma,\nu}(\cdot,y)$ as well as $tD_t\mathbf{N}e_{\gamma,\nu}(\cdot,y)$ are functions in $\mathcal{C}^1$, then

$$v_{\gamma,\nu}(y) := \frac{2}{\rho}\, t\, D_t\, \mathbf{N}e_{\gamma,\nu}(\cdot,y) \tag{17.13}$$

is a solution of $-\mathbf{N}^*\, D_t v_{\gamma,\nu}(y) = e_{\gamma,\nu}(\cdot,y)$. As mentioned in Remark 17.3, we may approximate $\mathbf{P}^* v_{\gamma,\nu}(y)$ by $-\mathbf{N}^*\, D_t v_{\gamma,\nu}(y)$ if $\tau \ll 1$ and thus can consider $v_{\gamma,\nu}(y)$ as a reconstruction kernel for the operator $\mathbf{P}$ associated with the mollifier $e_{\gamma,\nu}(\cdot,y)$.

In order to compute (17.13), we have to calculate $D_t\mathbf{N}f$ for a function f which is a translation of a rotationally symmetric function. The steps of that calculation are outlined in Lemma 17.7 and Theorem 17.8.

Lemma 17.7. *Let $\varphi : [0,\infty) \to \mathbb{R}$ be continuous and Φ be an antiderivative of φ. Assume $y \in \mathbb{R}^3$ and define $f_\varphi \in \mathcal{C}(\mathbb{R}^3)$ by $f_\varphi(x) := \varphi(\|x-y\|^2)$. Then*

$$(\mathbf{M}f_\varphi)(x,t) = \begin{cases} \frac{\Phi\left((\|x-y\|+t)^2\right)-\Phi\left((\|x-y\|-t)^2\right)}{4\,t\,\|x-y\|}\,, & \text{if } x \neq y\,,\\ \varphi(t^2)\,, & \text{if } x = y\,. \end{cases} \tag{17.14}$$

for positive t, $x \in \mathbb{R}^3$ and $\mathbf{M}f_\varphi(x,0) = \varphi(\|x-y\|^2) = f_\varphi(x)$.

Proof. The identity $\mathbf{M}f_\varphi(x,0) = \varphi(\|x-y\|^2)$ immediately follows from the definitions of f_φ and $\mathbf{M}$, see (17.5). Let $t > 0$ and $x \in \mathbb{R}^3$. If $x = y$, then $f_\varphi(x+t\omega)$ is constant on $\omega \in S^2$ and hence $\mathbf{M}f_\varphi(x,t) = \varphi(t^2)$. If $x \neq y$, then

$$\begin{aligned} \mathbf{M}f_\varphi(x,t) &= \frac{1}{4\pi}\int_{S^2} \varphi(\|x+t\omega-y\|^2)\,\mathrm{d}S_2(\omega)\\ &= \frac{1}{4\pi}\int_{S^2} \varphi\left(\|x-y\|^2 + t^2 + 2t\,\|x-y\|\left\langle \frac{x-y}{\|x-y\|},\omega\right\rangle\right)\mathrm{d}S_2(\omega)\,. \end{aligned}$$

To evaluate the last integral, we apply the Funck-Hecke theorem for $n = 3$, see e.g. [78, p. 20], leading to

$$\mathbf{M}f_\varphi(x,t) = \frac{1}{2}\int_{-1}^{1} \varphi(\|x-y\|^2 + t^2 + 2t\,\|x-y\|\,s)\,ds\,.$$

Since Φ is an antiderivative of φ we find

$$\begin{aligned} \mathbf{M}f_\varphi(x,t) &= \frac{\Phi(\|x-y\|^2 + t^2 + 2t\,\|x-y\|\,s)\big|_{-1}^{1}}{4t\,\|x-y\|}\\ &= \frac{\Phi\left((\|x-y\|+t)^2\right) - \Phi\left((\|x-y\|-t)^2\right)}{4t\,\|x-y\|}\,, \end{aligned}$$

which proves (17.14). □

Theorem 17.8. *Let φ, Φ, and f_φ be as in Lemma 17.7 and assume φ to be a $\mathcal{C}^1$-function such that there is some $0 < \gamma < 1$ with supp $\varphi \subset [0, \gamma^2]$. Furthermore, assume that $y \in B_{\rho-\gamma}$. Then $f_\varphi \in \mathcal{C}_0^1(B_\rho)$,*

$$D_t \mathbf{N} f_\varphi \in \mathcal{C}_0^1(S_\rho \times [0, 2\rho])$$

and

$$(D_t \mathbf{N} f_\varphi)(\sigma, t) = p_\varphi(\|\sigma - y\|, t)\,, \qquad \sigma \in S_\rho\,, \quad t \in [0, 2\rho]\,, \tag{17.15}$$

where

$$p_\varphi(s, t) := \frac{(s-t)\varphi\left((s-t)^2\right)}{2\,s}\,, \qquad s > 0\,, \quad t \in [0, 2\rho]\,. \tag{17.16}$$

Proof. Since $\|\cdot\|^2 \in \mathcal{C}^\infty(\mathbb{R}^n)$, we have that $f_\varphi = \varphi \circ \|\cdot\|^2 \in \mathcal{C}^1(\mathbb{R}^3)$. To show that supp $f_\varphi \subset\subset B_\rho$ let $\eta := (\|y\| + \gamma + \rho)/2$ and $x \in \mathbb{R}^3 \backslash B_\eta$. From $\|y\| < \rho - \gamma$ we see that $0 < \eta < \rho$ and thus $B_\eta \subset\subset B_\rho$. An application of the triangle inequality yields $\|x - y\| \geq \|x\| - \|y\| \geq \eta - \|y\| = (\rho + \gamma - \|y\|)/2 > 0$ and hence $f_\varphi(x) = 0$ which implies $f_\varphi \in \mathcal{C}_0^1(B_\rho)$.
We aim now to prove (17.15). Let $t > 0$ and $\sigma \in S_\rho$. Since $(\mathbf{N} f_\varphi)(\sigma, t) = (t\,\mathbf{M} f_\varphi)(\sigma, t)$ and $\|\sigma - y\| > \gamma > 0$, we deduce from Lemma 17.7

$$(D_t \mathbf{N} f_\varphi)(\sigma, t) = \frac{\partial}{\partial t}\left(\frac{\Phi\left((\|\sigma - y\| + t)^2\right) - \Phi\left((\|\sigma - y\| - t)^2\right)}{4\,\|\sigma - y\|}\right)\,. \tag{17.17}$$

Since $\|\sigma - y\| + t > \gamma$, the first term at the right-hand side vanishes. Moreover, since $\Phi' = \varphi$, equation (17.17) implies (17.15). For $t = 0$ equation (17.15) holds true anyway, since both sides are equal to 0.
Finally, assertion $D_t \mathbf{N} f_\varphi \in \mathcal{C}_0^1(S_\rho \times [0, 2\rho])$ is an immediate consequence of (17.15) and (17.16). □

We are now able to compute an explicit representation of $v_{\gamma,\nu}(y)$.

Corollary 17.9. *Let $v_{\gamma,\nu}(y)$ be defined by (17.13) with $e_{\gamma,\nu}(\cdot, y)$ as in (17.10), (17.11) and let $\nu > 1$. Furthermore, assume $y \in B_{\rho-\gamma}$ and $0 < \gamma < 1$. Then we have the representation*

$$v_{\gamma,\nu}(y)(\sigma, t) = \frac{k_{\gamma,\nu}(\|\sigma - y\|, t)}{4\,\pi\,\|\sigma - y\|}\,, \qquad \sigma \in S_\rho\,, \quad t \in [0, 2\rho]\,, \tag{17.18}$$

where

$$k_{\gamma,\nu}(s, t) := \frac{4\,\pi\,t\,(s-t)\,\rho}{\gamma^3\,I_\nu}\,R_\nu\left((s-t)^2/\gamma^2\right) \tag{17.19}$$

and R_ν is defined as in (17.11).
Particularly, if $\|\sigma - y\| \notin [t - \gamma, t + \gamma]$, then $v_{\gamma,\nu}(y)(\sigma, t) = 0$.

Proof. The assumption $\nu > 1$ guarantees that $\varphi(r^2) = R_\nu(r^2/\gamma^2)/(I_\nu\,\gamma^3)$ and hence $f_\varphi = e_{\gamma,\nu}(\cdot, y)$ satisfies the requirements of Theorem 17.8. Hence, from (17.13), we deduce

$$\begin{aligned} \upsilon_{\gamma,\nu}(y)(\sigma,t) &= \frac{2}{\rho\,\gamma^3\,I_\nu}\frac{t}{}\,\frac{(\|\sigma-y\|-t)\,R_\nu\left((\|\sigma-y\|-t)^2/\gamma^2\right)}{2\,\|\sigma-y\|} \\ &= \frac{1}{4\,\pi\,\|\sigma-y\|}\,\frac{4\,\pi\,t\,(\|\sigma-y\|-t)}{\rho\,\gamma^3\,I_\nu}\,R_\nu\left((\|\sigma-y\|-t)^2/\gamma^2\right), \end{aligned}$$

whence (17.18), (17.19) follow. From the fact that supp $R_\nu = [0,1]$, we find $\upsilon_{\gamma,\nu}(y)(\sigma,t) = 0$ for $\|\sigma - y\| \notin [t-\gamma, t+\gamma]$. □

A plot of $\upsilon_{\gamma,\nu}(0)(\sigma,t)$ for $\rho = 1$ can be found in the picture to the right in Figure 17.3.

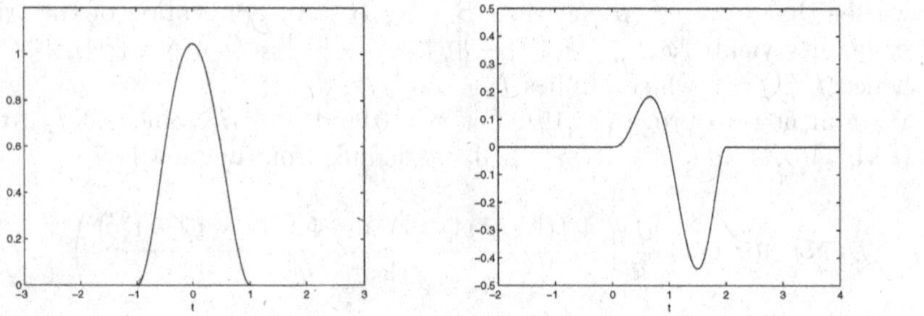

Fig. 17.3. Plots of the radial part $e_{\gamma,\nu}(t,0)$, $t = \|x\|$, of the mollifier $e_{\gamma,\nu}(\cdot,0)$ (17.10, left picture) and of the associated reconstruction kernel $\upsilon_{\gamma,\nu}(0)(\|\sigma\| = 1,t)$ (17.18, right picture) for $\gamma = 1$, $\nu = 2$ and $\rho = 1$. Both functions are compactly supported. The support of $e_{\gamma,\nu}(t,0)$ is $[-\gamma,\gamma] = [-1,1]$ whereas the support of $\upsilon_{\gamma,\nu}(0)(\|\sigma\| = 1,t)$ is given by the closed interval $[-\gamma+1,\gamma+1] = [0,2]$. The graph of the reconstruction kernel $\upsilon_{\gamma,\nu}$ clearly illuminates the differentiation D_t involved in (17.17) as well as the shift by $\|\sigma\| = 1$, see representation (17.18).

Let $\gamma > 0$, $\nu > 1$ be fixed and assume $y \in B_{\rho-\gamma}$ and define $p := \mathbf{P}f$. Taking into account that $\mathbf{P}^*\upsilon_\gamma(y) = -\mathbf{N}\,D_t\upsilon_\gamma(y)$, if j is replaced by the delta distribution, we may consider (17.18) as an appropriate choice for the reconstruction kernel associated with the mollifier $e_{\gamma,\nu}$. Note that in this case we set $\tau = 0$. Hence, assuming a delta function pulse j, the method of approximate inverse applied to $\mathbf{P}f = p$ then reads as

$$\begin{aligned} f_{\gamma,\nu}(y) &= \widetilde{\mathbf{P}}_{\gamma,\nu}\,\mathbf{P}f = \langle \mathbf{P}f, \upsilon_{\gamma,\nu}(y)\rangle_{\rho,2\rho} \\ &= \int_{S_\rho}\int_0^{2\rho} p(\sigma,t)\,\frac{k_{\gamma,\nu}(\|\sigma-y\|,t)}{4\,\pi\,\|\sigma-y\|}\,\mathrm{d}t\,\mathrm{d}S_2^\rho(\sigma) \end{aligned}$$

$$= \int_{S_\rho} \frac{1}{4\pi\,\|\sigma - y\|} \left(\int_0^{2\rho} p(\sigma,t)\, k_{\gamma,\nu}(\|\sigma - y\|, t)\, \mathrm{d}t \right) \mathrm{d}S_2^\rho(\sigma)$$

$$= (\mathbf{N}^* q_{\gamma,\nu})(y)$$

with

$$q_{\gamma,\nu}(\sigma, s) := \int_{s-\gamma}^{s+\gamma} p(\sigma,t)\, k_{\gamma,\nu}(s,t)\, \mathrm{d}t \qquad (17.20)$$

which represents an inversion scheme of *filtered backprojection type.*

Remark 17.10. The assumption $y \in B_{\rho-\gamma}$ is not a significant restriction with respect to applications, since $0 < \gamma \ll 1$ and the support of f has in fact a positive distance from $\partial B_\rho = S_\rho$ in practical experiments.

17.3 Numerical results

We conclude this chapter to show the performance of the method by some numerical experiments where we again use a semi-discrete setting. We want to recover a function f from measurement data $p = \mathbf{P}f$. Let $\gamma > 0$ and $\nu > 1$ be fixed positive numbers. The approximations $f_{\gamma,\nu}(y) := \langle p, v_{\gamma,\nu}(y)\rangle_{\rho,2\rho}$ consist of first computing the filtered signal $q_{\gamma,\nu}$ defined by (17.20) followed by the evaluation of the backprojection

$$f_{\gamma,\nu}(y) = (\mathbf{N}^* q_{\gamma,\nu})(y) = \frac{1}{4\pi} \int_{S_\rho} \frac{q_{\gamma,\nu}(\sigma, \|y - \sigma\|)}{\|y - \sigma\|}\, \mathrm{d}S_2^\rho(\sigma)$$

in every reconstruction point $y \in B_\rho$. Hence, the algorithm consists of two steps: First we perform a filtering step and then we integrate over all spheres with center on S_ρ intersecting y. For our numerical tests we set $\rho = 1$, that is $S_\rho = S^2$. We furthermore assume that the data are merely known for a finite number of $N_\theta N_\varphi$ detector points

$$\sigma_{k,l} = \begin{pmatrix} \cos(\theta_k)\cos(\phi_l) \\ \cos(\theta_k)\sin(\phi_l) \\ \sin(\theta_k) \end{pmatrix} \in S^2 = S_\rho\,, \quad k = 1, \ldots, N_\theta\,, \quad l = 1, \ldots, N_\phi\,,$$

with $\theta_k := -\pi/2 + \pi\,(k-1)/(N_\theta - 1)$ and $\phi_l := 2\pi(l-1)/N_\phi$, and the pressure signal at each detector point is sampled at N_t time steps

$$t_m = 2\,(m-1)/N_t\,, \quad m = 1, \ldots, N_t\,.$$

The aim is to evaluate $f_{\gamma,\nu}$ at $N := N_y^3$ points y_i, $i = 1, \ldots, N$, located at an equidistant mesh grid. This requires the computation of $q_{\gamma,\nu}(\sigma_{k,l}, \|\sigma_{k,l} - y_i\|)$ in every reconstruction point y_i. To reduce the computational effort, we evaluate

$q_{\gamma,\nu}(\sigma_{k,l},\cdot)$ for t_m with $m = 1,\ldots,N_t$ only and use linear interpolation to approximately find the value at $\|\sigma_{k,l} - y_i\|$. As quadrature rule on S^2 for a function F on the sphere, we use the trapezoidal rule in θ and ϕ applied to the coordinate representation $\cos(\theta)\,F(\theta,\phi)$.

Let us assume that $O(N_\theta) = O(N_\phi) = O(N_t) = O(N_y)$. The total number N_{op} of operations needed to perform this algorithm computes as

$$N_{\mathrm{op}} = O(N_t^2) + O(N_\theta\, N_\phi)\Big(O(N_t^2) + O(N_y^3)\Big) = O(N_y^5) = O(N^{5/3}) .$$

We show reconstruction results for two different examples.

1. Let us consider an energy deposition function $f \in C_0^1(B_\rho)$ of the form

$$f(y) = \sum_{\alpha=1}^{M} F_\alpha(\|y - y_\alpha\|) ,$$

consisting of M radially symmetric absorbers $F_\alpha(\|y - y_\alpha\|)$ with centers y_α and radial profiles F_α. From Theorem 17.8, we compute data $p = \sum_\alpha p_\alpha$ with

$$p_\alpha(\sigma, t) = \frac{\|\sigma - y_\alpha\| - t}{2\,\|\sigma - y_\alpha\|}\, F_\alpha\Big(\big|\|\sigma - y_\alpha\| - t\big|\Big)$$

for $\sigma \in S^2$ and $t \geq 0$. To produce the results of Figure 17.5, we used an object consisting of $M = 8$ balls of different radii, centers and densities. Hence, $F_\alpha(z) = d_\alpha\, \chi_{[0,r_\alpha]}(z)$ are characteristic functions of closed intervals $[0, r_\alpha]$ where r_α denotes the radius and d_α the density of ball α. The centers are given by y_α. The table of Figure 17.4 shows the parameters which were used in the reconstructions of Figure 17.5.

α	r_α	d_α	y_α
1	0.8	4	(-0.1, 0, 0)
2	0.7	-4	(-0.1, 0, 0)
3	0.2	2	(-0.4, 0, 0)
4	0.1	2	(-0.4, 0, 0)
5	0.1	2.5	(-0.05, 0, 0)
6	0.25	2	(0.4, 0, 0)
7	0.2	-2	(0.4, 0, 0)
8	0.1	4.5	(0.4, 0, 0)

Fig. 17.4. Parameters of the phantom used for the reconstructions in Figure 17.5.

We have displayed a cross section through the plane $\{y_3 = 0\}$ and computed reconstructions on the grid $(y_1^m, y_2^{m'}, 0) \subset [-1,1] \times \{0\}$, where

$$y_1^m = -1 + 2\,(m-1)/N_y\,, \quad y_2^{m'} = -1 + 2\,(m'-1)/N_y\,, \quad 1 \leq m, m' \leq N_y\,.$$

The algorithm has been performed with $N_y = N_t = N_\phi = 120$ and $N_\theta = 60$. Figure 17.5 shows both the reconstruction of the exact data as well as of noisy data with a random perturbation of 20% additive Gaussian noise. The regularization parameter γ has been chosen to be 0.05 and the exponent in the mollifier was $\nu = 2$.

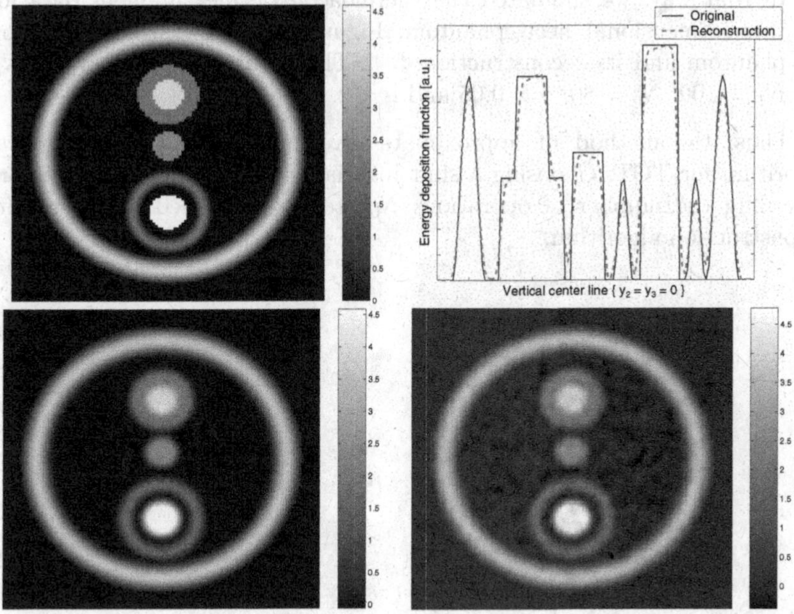

Fig. 17.5. Cross section $y_3 = 0$ of the object from which the thermoacoustic pressure is calculated analytically (upper left picture). Vertical centerline of the original and the reconstruction from exact data (upper right picture). Reconstruction from exact data (lower left picture). Reconstruction from noisy data perturbed with 20% Gaussian white noise (lower right picture). The reconstructions were computed on $(y_1, y_2) \in [-1, 1]^2$ using an equidistant mesh grid with N_y^2 grid points, $N_y = 120$.

'The picture was reprinted from M. Haltmeier, T. Schuster, and O. Scherzer, *Filtered backprojection for thermoacoustic computed tomography in spherical geometry*, Math. Meth. Appl. Sci., 28 (2005), pp. 1919–1937. Copyright ©2005 John Wiley & Sons Limited. Reproduced with permission.'

2. Consider an arbitrary energy deposition function $f \in C_0^1(B_\rho)$. To simulate the measurement data, we have to find $p(\sigma, t)$ for $\sigma \in S^2$ and $t \in [0, 2]$ numerically. To this end, we use Fourier series expansions

$$f(y) = \frac{1}{8} \sum_{k \in \mathbb{Z}^3} f_k \, \mathrm{e}^{-\mathrm{i}\,\pi\,\langle k,y \rangle/2} \tag{17.21}$$

and

$$p(y,t) = \frac{1}{8} \sum_{k \in \mathbb{Z}^3} p_k(t) \, \mathrm{e}^{-\mathrm{i}\,\pi\,\langle k,y \rangle/2} \tag{17.22}$$

for $y \in [-2, 2]^3$ and $t \geq 0$. Defining

$$p_k(t) = \cos(\pi \, \|k\| \, t/2) \, f_k \,, \tag{17.23}$$

then p agrees with the unique solution of (17.2), (17.3) on $S^2 \times [0, 2]$ (for $\tau \to 0$). By means of (17.21), (17.22), (17.23), we compute approximations to $p = D_t \mathbf{N} f$ using the fast Fourier transform.
In that way, we simulated the thermoacoustic measurement data for a three-dimensional head phantom. Figure 17.6 shows the Shepp-Logan phantom and its reconstruction $f_{\gamma,\nu}$. The parameters were $N_y = N_t = N_\phi = 100$, $N_\theta = 80$, $\gamma = 0.05$ and $\nu = 2$.

Thus, the method of approximate inverse delivers a stable inversion algorithm for TCT. Choosing a shift invariant mollifier $e_{\gamma,\nu}$ saves one order regarding the number of operations N_{op} to be performed in the resulting reconstruction algorithm.

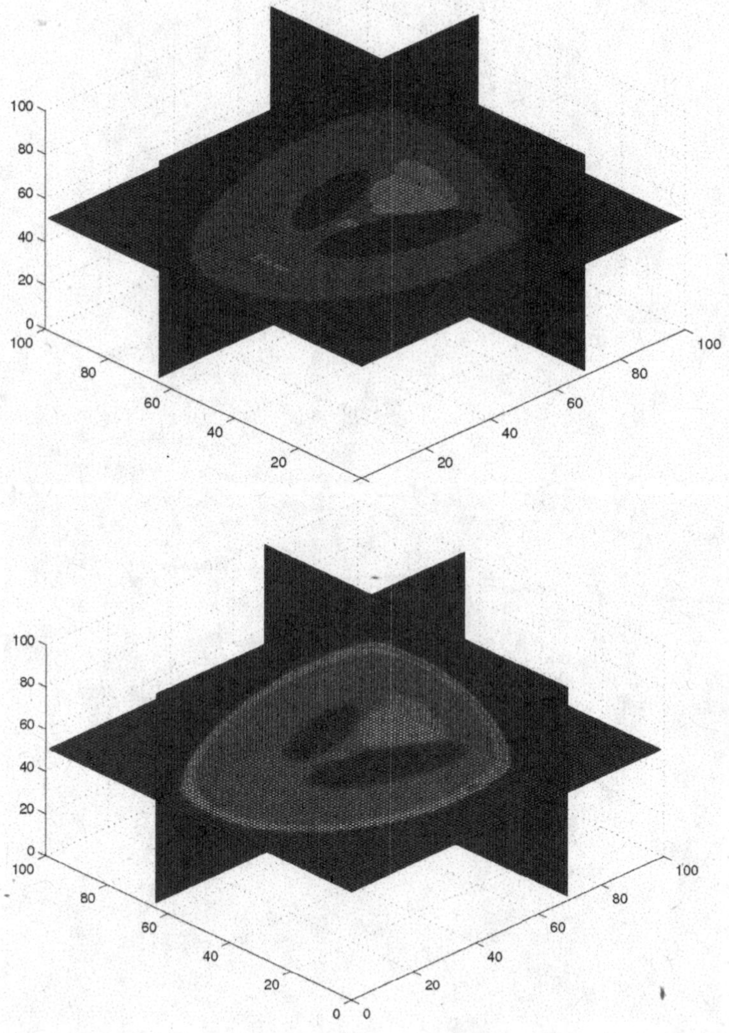

Fig. 17.6. Upper picture: original head phantom. Lower picture: Reconstruction from simulated data with $\gamma = 0.05$ and $\nu = 2$. The reconstructions were computed in an equidistant mesh grid in $[-2, 2]^3$.

18

Computation of reconstruction kernels in 3D computerized tomography

The material of this chapter is based on LOUIS [70]. We already dealt with the problem of 2D computerized tomography (CT) in Section 2.2. 3D - CT is concerned with the problem of recovering a three-dimensional quantity f from X-ray measurements where the X-ray sources are located on a curve $\Gamma \subset (\mathbb{R}^3 \backslash \Omega)$ and $\Omega \subset \mathbb{R}^3$ denotes the object under consideration. The problem is described by the cone beam transform

$$\mathbf{X}f(a,\omega) = \int_0^\infty f(a + t\omega)\,\mathrm{d}t\,,^1 \tag{18.1}$$

where $a \in \Gamma$ is the source from which the X-rays are emitted and $\omega \in S^2$ means a normalized vector of direction of the X-ray beam. Common source curves are e.g. a circle, two circles which are perpendicular to each other or a helix. We always assume that the function f to be recovered has its support in a bounded domain $\Omega \subset \mathbb{R}^3$ and make all further considerations for $\Omega = \Omega^3$, the open unit ball in $\mathbb{R}^3$. Figure 18.1 illustrates the situation where Γ consists of two perpendicular circles surrounding the object Ω^3.

We first show that $\mathbf{X}$ is a bounded mapping between appropriate L^2-spaces, if the scanning curve Γ satisfies a certain requirement.

Theorem 18.1. *Let $a \in \Gamma$. The mappings $\mathbf{X}_a : L^2(\Omega^3) \to L^2(S^2)$ defined by $\mathbf{X}_a f(\omega) = \mathbf{X}f(a,\omega)$ and $\mathbf{X} : L^2(\Omega^3) \to L^2(\Gamma \times S^2)$ are bounded, if*

$$\int_\Gamma (\|a\| - 1)^{-2}\,\mathrm{d}a < \infty\,. \tag{18.2}$$

Proof. For $f \in L^2(\Omega^3)$ and a source point $a \in \Gamma$ we have

[1] Although in relevant literature the cone beam transform usually is denoted by $\mathbf{D}$, we are using $\mathbf{X}$ to avoid confusions with the Doppler transform.

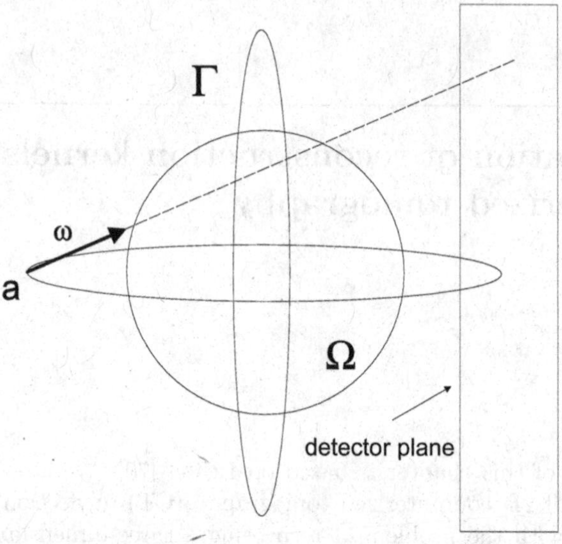

Fig. 18.1. Scanning geometry using two circles which are perpendicular to each other as scanning curve Γ.

$$\int_{S^2} |\mathbf{X}_a f(\omega)|^2 \, \mathrm{d}\omega = \int_{S^2} \Big| \int_0^\infty f(a + t\,\omega)\, \mathrm{d}t \Big|^2 \mathrm{d}\omega \le 2 \int_{S^2} \int_0^\infty |f(a + t\,\omega)|^2 \, \mathrm{d}t \, \mathrm{d}\omega$$
$$= 2 \int_{\Omega^3} |f(x)|^2 \, \|x - a\|^{-2} \, \mathrm{d}x \le 2\,(\|a\| - 1)^{-2} \, \|f\|^2_{L^2(\Omega^3)} \,,$$

where we used the substitution $x = a + t\,\omega$ and the fact that $f(x) = 0$ in $\mathbb{R}^3 \backslash \overline{\Omega^3}$. This shows the continuity of $\mathbf{X}_a$. The continuity of $\mathbf{X}$ follows then by using $\mathbf{X}(a, \omega) = \mathbf{X}_a(\omega)$ and

$$\int_\Gamma \int_{S^2} |\mathbf{X} f(a, \omega)|^2 \, \mathrm{d}\omega \, \mathrm{d}a \le 2 \, \|f\|^2_{L^2(\Omega^3)} \int_\Gamma (\|a\| - 1)^{-2} \, \mathrm{d}a \,.$$

□

Theorem 18.1 implies that $\mathbf{X}_a$ and $\mathbf{X}$ have linear and bounded adjoints $\mathbf{X}_a^*$, $\mathbf{X}^*$.

Lemma 18.2. *The adjoints* $\mathbf{X}_a^* : L^2(S^2) \to L^2(\Omega^3)$ *and* $\mathbf{X}^* : L^2(\Gamma \times S^2) \to L^2(\Omega^3)$ *have representations*

$$\mathbf{X}_a^* g(x) = \|x - a\|^{-2} \, g\Big(\frac{x - a}{\|x - a\|} \Big) \,, \tag{18.3a}$$

$$\mathbf{X}^* g(x) = \int_\Gamma \|x - a\|^{-2} \, g\Big(\frac{x - a}{\|x - a\|} \Big) \, \mathrm{d}a \,. \tag{18.3b}$$

Proof. Let $f \in L^2(\Omega^3)$, $g \in L^2(S^2)$. Then

$$\begin{aligned}\int_{S^2} \mathbf{X}_a f(\omega)\, g(\omega)\, \mathrm{d}\omega &= \int_{S^2}\int_0^\infty f(a + t\,\omega)\, g(\omega)\, \mathrm{d}t\, \mathrm{d}\omega \\ &= \int_{\Omega^3} \|x - a\|^{-2} f(x)\, g((x-a)/\|x-a\|)\, \mathrm{d}x \\ &= \langle f, \mathbf{X}_a^* g\rangle_{L^2(\Omega^3)}\,.\end{aligned}$$

Here again, we substituted $x = a + t\,\omega$. This shows (18.3a). Representation (18.3b) follows easily from (18.3a) by an integration over Γ. □

To compute reconstruction kernels for $\mathbf{X}$ we have to solve equations

$$\mathbf{X}^* v_\gamma(x) = e_\gamma(x,\cdot)\,, \tag{18.4}$$

where $e_\gamma(x,y)$ is a mollifier in the sense of definition 2.1. To cope with equations as (18.4), an appropriate inversion formula for $\mathbf{X}$ might be useful. LOUIS outlines in [69, 70] how to deduce such an inversion formula and how to get reconstruction kernels using the *formula of Grangeat.* Grangeat's formula is based on a more general identity obtained by HAMAKER ET AL. [39] and states a connection between the cone beam transform $\mathbf{X}$ and the three-dimensional Radon transform $\mathbf{R}$. We have

$$\begin{aligned}\frac{\partial}{\partial s}\mathbf{R} f(\omega, s = \langle a,\omega\rangle) &= -\int_{S^2} \mathbf{X} f(a,\theta)\, \delta'(\langle\theta,\omega\rangle)\, \mathrm{d}\theta && (18.5\mathrm{a})\\ &= \int_{S^2\cap\{\langle\theta,\omega\rangle=0\}} \nabla_y \mathbf{X} f(a, y=\theta)\, \mathrm{d}S_1(\theta)\,. && (18.5\mathrm{b})\end{aligned}$$

Formula (18.5a) has been proven by GRANGEAT, see [33, 34]. Simple proofs can be found in LOUIS [70] and NATTERER, WÜBBELING [84]. Identity (18.5b) explains how the derivative of the delta distribution δ' is to be understood. Note that the Radon transform $\mathbf{R}$ in (18.5a) integrates over planes in $\mathbb{R}^3$. Louis [70] takes the inversion formula of the 3D Radon transform as starting point

$$f(x) = -\frac{1}{8\pi^2}\int_{S^2} \frac{\partial^2}{\partial s^2}\mathbf{R} f(\omega, \langle x,\omega\rangle)\, \mathrm{d}\omega\,,$$

which can be written also as

$$f(x) = \frac{1}{8\pi^2}\int_{S^2}\int_{\mathbb{R}} \frac{\partial}{\partial s}\mathbf{R} f(\omega, s)\, \delta'(s - \langle x,\omega\rangle)\, \mathrm{d}s\, \mathrm{d}\omega\,. \tag{18.6}$$

KIRILLOV [58] and TUY [126] showed that full reconstruction from cone-beam data is possible, if the source curve Γ intersects each plane which passes the support of f transversally. Assume $a = a(\alpha) : I \to \mathbb{R}^3$ to be a parameterization of the scanning curve $\Gamma = \mathsf{R}(a)$, then this condition means that for each $x \in \operatorname{supp} f$ and direction $\omega \in S^2$ there exists an $\alpha = \alpha(x,\omega)$ satisfying

$$\langle a(\alpha), x\rangle = \langle \omega, x\rangle\,, \qquad \langle a'(\alpha), \omega\rangle \neq 0\,. \tag{18.7}$$

Condition (18.7) is known as *Tuy-Kirillov condition* . If we further denote by $n = n(\omega, s)$ the number of source points $a \in \Gamma$ with $s = \langle a, \omega\rangle = \langle x, \omega\rangle$ and $m = 1/n$, which is possible since $n \geq 1$ by the Tuy-Kirillov condition, then we may substitute $s = \langle a, \omega\rangle$ in (18.6) and apply Grangeat's formula (18.5a) to obtain

$$\begin{aligned} f(x) &= \frac{1}{8\pi^2}\int_{S^2}\int_\Gamma \frac{\partial}{\partial s}\mathbf{R}f(\omega, \langle a, \omega\rangle)\,\delta'(\langle a - x, \omega\rangle)\,|\langle a', \omega\rangle|\,m(\omega, \langle a, \omega\rangle)\,\mathrm{d}a\,\mathrm{d}\omega \\ &= -\frac{1}{8\pi^2}\int_{S^2}\int_\Gamma\Big(\int_{S^2}\mathbf{X}f(a, \theta)\,\delta'(\langle \theta, \omega\rangle)\,\mathrm{d}\theta \\ &\qquad\qquad \times\delta'(\langle a - x, \omega\rangle)\,|\langle a', \omega\rangle|\,m(\omega, \langle a, \omega\rangle)\Big)\,\mathrm{d}a\,\mathrm{d}\omega \\ &= -\frac{1}{8\pi^2}\int_\Gamma \|x - a\|^{-2}\int_{S^2}\Big(\int_{S^2}\mathbf{X}f(a, \theta)\,\delta'(\langle \theta, \omega\rangle)\,\mathrm{d}\theta \\ &\qquad\qquad \times\delta'(\langle (x - a)/\|x - a\|, \omega\rangle)\,|\langle a', \omega\rangle|\,m(\omega, \langle a, \omega\rangle)\Big)\,\mathrm{d}\omega\,\mathrm{d}a\,. \end{aligned}$$

In the last step, we used that δ' is homogeneous of degree -2. The number $m(\omega, s) \in \mathbb{N} \cup \{0\}$ is called *Crofton symbol* and equals the number of intersection points of the plane $\{x \in \mathbb{R}^3 : \langle x, \omega\rangle = s\}$ with the source trajectory Γ. Introducing operators

$$Tg(a, \omega) = \int_{S^2} g(a, \theta)\,\delta'(\langle \theta, \omega\rangle)\,\mathrm{d}\theta\,, \qquad a \in \Gamma\,, \quad \omega \in S^2$$

and

$$M_{\Gamma,a}h(a, \omega) = |\langle a', \omega\rangle|\,m(\omega, \langle a, \omega\rangle)\,h(a, \omega)\,, \qquad a \in \Gamma\,, \quad \omega \in S^2$$

we may summarize the results as stated in the following theorem.

Theorem 18.3 (Louis [**70**]). *Assume that the Tuy-Kirillov condition and* (18.2) *are satisfied. Then the inversion formula for the cone beam transform has a representation*

$$f = -\frac{1}{8\pi^2}\mathbf{X}^* T\,M_{\Gamma,a}\,T\,\mathbf{X}f\,. \tag{18.8}$$

Relying on formula (18.8), we are able to compute reconstruction kernels for $\mathbf{X}$. If a mollifier $e_\gamma(x, y) \in L^2(\mathbb{R}^3, \mathbb{R}^3)$ is given, then an associated reconstruction kernel $\upsilon_\gamma(x) \in L^2(\Gamma \times S^2)$ can be computed as

$$\upsilon_\gamma(x) = -\frac{1}{8\pi^2}T\,M_{\Gamma,a}\,T\,\mathbf{X}e_\gamma(x, \cdot)\,, \qquad x \in \Omega^3\,. \tag{18.9}$$

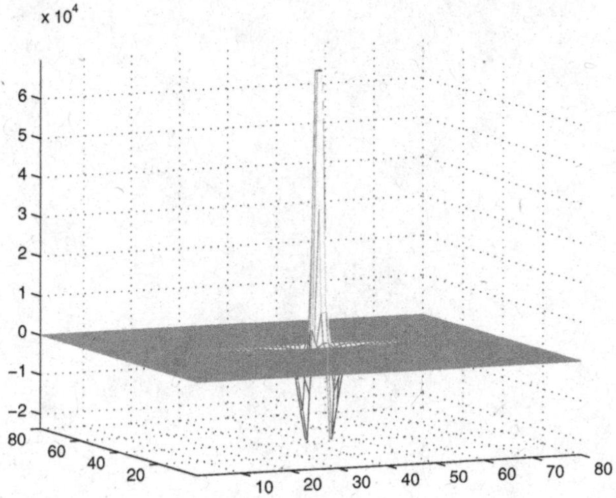

Fig. 18.2. Reconstruction kernel for the cone beam transform **X**.

A reconstruction kernel which has been computed according to formula (18.9) is displayed in Figure 18.2.[2]

Remark 18.4. The operators $\mathbf{X}^*$ and $M_{\Gamma,a}$ depend on the scanning curve Γ and so do any invariance properties which might decrease the cost of computing reconstruction kernels. A detailed analysis of the usage of invariances in practical applications of the approximate inverse in 3D X-ray tomography with a circular scanning geometry can be found in LOUIS [69]. A further inversion formula for the special case when Γ equals a spiral was found by KATSEVICH [53], see also NOO, PACK, AND HEUSCHER [87]. The possibility of calculating reconstruction kernels using Katsevich's formula is subject of current research. We also refer to KATSEVICH [54, 55], NOO, DEFRISE, AND KUDO [86], WANG ET AL. [128] for more general trajectories, and KATSEVICH [56] for cone beam local tomography.

[2] Courtesy of Prof. Dr. A.K Louis, Department of Mathematics, Saarland University, 66041 Saarbrücken, Germany. This picture has been taken from his article *Filter design in three-dimensional cone beam tomography: circular scanning geometry*, Inverse Problems, 19 (2003), pp. S31–S40. Copyright ©2003 IOP Publishing Limited. Reprinted with permission.'

19

Conclusion and perspectives

We outlined in the last part of the book that there are various possible applications of the method of approximate inverse in industry and medical imaging. In some cases the method did not only lead to novel and very efficient solvers but allowed also for a detailed convergence analysis. Nevertheless, there are still more applications which have not been discussed in this book at all. A large area of current research are scattering problems. The first publication concerning a reconstruction method for inverse scattering using the method of approximate inverse is ABDULLAH, LOUIS [1]. Here, the authors considered the *Lippmann-Schwinger equation*

$$u(\alpha, x) = u_{\text{inc}}(\alpha, x) - k^2 \int_{\|y\|<R} G(k\,\|x-y\|)\, f(y)\, u(\alpha, y)\, \mathrm{d}y\,, \qquad (19.1)$$

where the entire field u is the sum of the scattered field and the incident field, $u = u_{\text{sc}} + u_{\text{inc}}$, k is the wave number, $f(x) = n^2(x) - 1$ with the refractive index n and

$$G(k\,\|x-y\|) = \frac{\mathrm{i}}{4}\, H_0^{(1)}(k\,\|x-y\|)\,, \qquad x \neq y$$

denotes the Green function, $H_0^{(1)}$ is the Hankel function of first kind and order 0. Solving equation (19.1) is equivalent to the inverse problem of recovering the refractive index n from measurements of the scattered field u_{sc}. In [1] this problem is solved by computing reconstruction kernels with the help of the singular value decomposition of the integral operator in (19.1). Improvements and extensions of the method, e.g. for solving three-dimensional problems related to Maxwell's equations, are still under consideration.

Inverse problems will play a role of increasing importance when dealing with questions in industry, natural science and medical imaging and the method of approximate inverse might be a powerful and important tool for finding new ways to cope with these problems.

References

1. H. Abdullah and A. Louis, *The Approximate Inverse for Solving an Inverse Scattering Problem for Acoustic Waves in an Inhomogeneous Medium*, Inverse Problems, 15 (1999), pp. 1213–1230.
2. M. Abramowitz and I. Stegun, *Handbook of mathematical functions*, Dover, New York, 1972.
3. R. A. Adams, *Sobolev Spaces*, Academic Press, New York, 1975.
4. F. Andersson, *The Doppler moment transform in Doppler tomography*, Inverse Problems, 21 (2005), pp. 1249–1274.
5. L.-E. Andersson, *On the determination of a function from spherical averages*, SIAM J. Math. Anal., 19 (1988), pp. 214–232.
6. J.-P. Aubin, *Applied functional analysis*, Wiley, Chichester, second ed., 1999.
7. M. Bebendorf and S. Rjasanow, *Matrix compression for the radiation heat transfer in exhaust pipes*, in Multifield Problems. State of the Art, W. Wendland, A.-M. Sändig, and W. Schiehlen, eds., Springer, 2000, pp. 183–192.
8. ——, *Adaptive low-rank approximation of collocation matrices*, Computing, 70 (2003), pp. 1–24.
9. M. Bezuglova, E. Derevtsov, and S. Sorokin, *The reconstruction of a vector field by finite difference methods*, J. Inv. Ill-Posed Problems, 10 (2002), pp. 125–154.
10. B. Borden and M. Cheney, *Synthetic-aperture imaging from high-Doppler-resolution measurements*, Inverse Problems, 21 (2005), pp. 1–11.
11. J. Bramble and S. Hilbert, *Estimation of linear functionals on Sobolev spaces with application to Fourier transforms and spline interpolation*, SIAM J. Numer. Anal., 7 (1970), pp. 112–124.
12. H. Braun and A. Hauck, *Tomographic reconstruction of vector fields*, IEEE Trans. Signal Proc., 39 (1991), pp. 464–471.
13. A. Bukgheim and S. Kazantsev, *Inversion formula for the fan-beam attenuated Radon transform in a unit disk.* Preprint No. 99, The Sobolev Institute of Mathematics of SB RAS, 2002.
14. G. Chen and J. Zhou, *Boundary Element Methods*, Academic Press, Cambridge, 1992.
15. M. Cheney, *Tomography problems arising in Synthetic Aperture Radar*, Contemporary Mathematics, 278 (2001), pp. 15–28.

16. A. Chorin and J. Marsden, *A Mathematical Introduction to Fluid Mechanics*, Springer, New York, 3rd ed., 1993.
17. F. Constantinescu, *Distributionen und ihre Anwendung in der Physik*, Teubner, 1974.
18. R. Courant and D. Hilbert, *Methods of Mathematical Physics, Vol. II*, Wiley-Interscience, New York, 1962.
19. A. Denisjuk, *Integral Geometry on the family of semi-spheres*, Fractional Calculus and Applied Analysis, 2 (1999), pp. 31–46.
20. A. Denisjuk, *On numerical reconstruction of a function by its arc means with incomplete data*, tech. report, Brest State University, Belarus, 1999.
21. A. Denisjuk, *Inversion of the X-ray transform for 3D symmetric tensor fields with sources on a curve*, Inverse Problems, 22 (2006), pp. 399–411.
22. E. Derevtsov and I. Kashina, *Numerical solution to the vector tomography problem by tools of a polynomial basis*, Sib. J. Numerical Math., 5 (2002), pp. 233–254. (in Russian).
23. E. Derevtsov, A. Louis, and T. Schuster, *Two approaches to the problem of defect correction in vector field tomography solving boundary value problems*, J. Inv. Ill-Posed Prob., 12 (2004), pp. 597–626.
24. L. Desbat, *Efficient parallel sampling in vector field tomography*, Inverse Problems, 11 (1995), pp. 995–1003.
25. B. Eigenmann, B. Scholtes, and E. Macherauch, *An improved technique for X-ray residual stress determinations on ceramics with steep subsurface stress gradients*, in Residual Stresses III, H. Fujiwara, T. Abe, and K. Tanaka, eds., Elsevier Applied Sciences, 1992, pp. 601–606.
26. H. Engl, M. Hanke, and A. Neubauer, *Regularization of Inverse Problems*, Kluwer, Dordrecht, Boston, London, 2000.
27. A. Faridani, D. Finch, E. Ritman, and K. Smith, *Local tomography II*, SIAM J. Appl. Math., 57 (1997), pp. 1095–1127.
28. A. Faridani and A. Rieder, *The semi-discrete filtered backprojection algorithm is optimal for tomographic inversion*, SIAM J. Numer. Anal., 41 (2003), pp. 869–892.
29. A. Faridani, E. Ritman, and K. Smith, *Local tomography*, SIAM J. Appl. Math., 52 (1992), pp. 459–484.
30. J. Fawcett, *Inversion of n-dimensional spherical averages*, SIAM J. Appl. Math., 45 (1985), pp. 336–341.
31. D. Finch, S. Patch, and Rakesh, *Determining a function from its mean values over a family of spheres*, SIAM J. Math. Anal., 35 (2004), pp. 1213–1240.
32. I. Gradshtein and I. Ryzhik, *Table of Integrals, Series and Products*, Academic Press, New York, sixth ed., 2000.
33. P. Grangeat, *Analyse d'un système d'imagerie 3D par reconstruction à partir de radiographie X en geometrie conique*, PhD thesis, École Normale Superieure des Télécommunication, Paris, France, 1987.
34. ———, *Mathematical framework of cone-beam reconstruction via the first derivative of the Radon transform*, in Lecture Notes in Math., G. Herman, A. Louis, and F. Natterer, eds., vol. 1497, New York, 1991, Springer, pp. 66–97.
35. V. Gusev and A. Karabutov, *Laser Optoacosutics*, Institute of Physics, New York, 1993.

36. M. Hahn and E. Quinto, *Distances between measures from 1-dimensional projections as implied by continuity of the inverse Radon transform*, Z. Wahr., 70 (1985), pp. 361–380.
37. M. Haltmeier, O. Scherzer, P. Burgholzer, and G. Paltauf, *Thermoacoustic computerized tomography with large plain receivers*, Inverse Problems, 20 (2004), pp. 1663–1673.
38. M. Haltmeier, T. Schuster, and O. Scherzer, *Filtered backprojection for thermoacoustic computed tomography in spherical geometry*, Math. Meth. Appl. Sci., 28 (2005), pp. 1919–1937.
39. C. Hamaker, K. Smith, D. Solmon, and S. Wagner, *The divergent beam X-ray transform*, Rocky Mountain J. Math., 10 (1980), pp. 252–283.
40. M. Hanke, *Accelerated Landweber iterations for the solution of ill-posed problems*, Numer. Math., 5 (1991), pp. 341–373.
41. ———, *Conjugate Gradient Type Methods for Ill-Posed Problems*, in Pitman Research Notes in Mathematics, vol. 327, Longman Scientific & Technical, Harlow, UK, 1995.
42. M. Hanke-Bourgeois, *Grundlagen der Numerischen Mathematik und des Wissenschaftlichen Rechnens*, Teubner, Stuttgart, 2002.
43. S. Helgason, *The Radon Transform*, Birkhäuser, Boston, 1980.
44. H. Helmholtz, *Über Integrale der Hydrodynamischen Gleichungen*, J. Reine Angew. Math., 55 (1858), pp. 25–55.
45. H.Hellsten and L.-E. Andersson, *An inverse method for the processing of synthetic aperture radar*, Inverse Problems, 3 (1987), pp. 111–124.
46. B. Hofmann, *Mathematik inverser Probleme*, Teubner, Stuttgart, Leipzig, 1999.
47. T. Jansson, M. Almqvist, K. Stråhlén, R. Eriksson, G. Sparr, H. Persson, and K. Lindström, *Ultrasound doppler vector tomography measurements of directional blood flow*, Ultrasound in Med. & Biol., 23 (1997), pp. 47–57.
48. F. John, *Partial Differential Equations*, Springer, Berlin, New York, 1982.
49. P. Jonas and A. Louis, *Approximate inverse for a one-dimensional inverse heat conduction problem*, Inverse Problems, 16 (2000), pp. 175–185.
50. P. Juhlin, *Principles of doppler tomography*, tech. report, Center for Mathematical Sciences, Lund Institute of Technology, SE-221 00 Lund, Schweden, 1992.
51. A. Kak and M. Slaney, *Principles of Computerized Tomographic Imaging*, IEEE Press, New York, 1988.
52. A. Kämpfe, *Röntgenographische Bestimmung von Texturen, Makro- und Mikroeigenspannungen an einphasigen Werkstoffen mit kubischer Kristallstruktur nach Kaltverformung*, Shaker Verlag, Aachen, 2001.
53. A. Katsevich, *Theoretically exact Filtered Back Projection-type inversion algorithm for spiral CT*, SIAM J. Appl. Math., 62 (2002), pp. 2012–2026.
54. ———, *A general scheme for constructing inversion algorithms for cone beam CT*, Int. J. Math. Sci., 21 (2003), pp. 1305–1321.
55. ———, *Image reconstruction for the circle and line trajectory*, Phys. Med. Biol., 49 (2004), pp. 5059–5072.
56. ———, *Improved cone beam local tomography*, Inverse Problems, 22 (2006), pp. 627–643.
57. S. Kazantsev and A. Bukgheim, *Singular value decomposition for the 2D fan-beam Radon transform of tensor fields*, J. Inv. Ill-Posed Prob., 12 (2004), pp. 245–278.

58. A. Kirillov, *On a problem of I.M. Gel'fand*, Soviet. Math. Dokl., 2 (1961), pp. 268–269.
59. J. Klein, *Rekonstruktionsverfahren für SAR: Inversion sphärischer Durchschnitte*, master's thesis, Westfälische Wilhelms - Universität, Institut für Numerische und Instrumentelle Mathematik, Münster, Germany, 2003.
60. R. Kruger, D. Reinecke, and G. Kruger, *Thermoacoustic Computed Tomography*, Medical Physics, 26 (1999), pp. 1832–1837.
61. L. Landau and E. Lifschitz, *Lehrbuch der theoretischen Physik, Band VI: Hydrodynamik*, Akademie Verlag, Berlin, 1991.
62. M. Lavrentiev, V. Romanov, and V. Vasiliev, *Multidimensiaonal inverse problems for differential equations*, vol. 167 of Lecture Notes in Mathematics, Springer Verlag, New York, 1970.
63. T. Leverenz, B. Eigenmann, and E. Macherauch, *Das Abschnitt - Polynom - Verfahren zur zerstörungsfreien Ermittlung gradientenbehafteter Eigenspannungszustände in den Randschichten bearbeiteter Keramiken*, Z. Metallkd., 87 (1996), pp. 616–625.
64. J. Lions and E. Magenes, *Non-Homogeneous Boundary Value Problems and Applications*, vol. 1, Springer, New York, 1972.
65. A. Louis, *Inverse und schlecht gestellte Probleme*, Teubner, Stuttgart, 1989.
66. ———, *Approximate inverse for linear and some nonlinear problems*, Inverse Problems, 12 (1996), pp. 175–190.
67. ———, *Application of the approximate inverse to 3D X-ray CT and ultrasound tomography*, in Inverse Problems in Medical Imaging and Nondestructive Testing, H. Engl, A. Louis, and W. Rundell, eds., Springer, Wien, New York, 1997, pp. 120–133.
68. ———, *A unified approach to regularization methods for linear ill-posed problems*, Inverse Problems, 15 (1999), pp. 489–498.
69. ———, *Filter design in three-dimensional cone beam tomography: circular scanning geometry*, Inverse Problems, 19 (2003), pp. S31–S40.
70. ———, *Development of algorithms in computerized tomography*, in The Radon Transform, Inverse Problems, and Tomography, G. Ólafsson and E. Quinto, eds., vol. 63 of Proceedings of Symposia in Applied Mathematics, AMS, 2006, pp. 25–42.
71. A. Louis and P. Maass, *A mollifier method for linear operator equations of the first kind*, Inverse Problems, 6 (1990), pp. 427–440.
72. A. Louis, P. Maass, and A. Rieder, *Wavelets*, Teubner, Stuttgart, 1994.
73. A. Louis and F. Natterer, *Mathematical problems in computerized tomography*, Proceedings IEEE, 71 (1983), pp. 379–389.
74. A. Louis and E. Quinto, *Local tomographic methods in SONAR*, in Surveys on solution methods for inverse problems, D. Colton, H. Engl, A. Louis, J. McLaughlin, and W. Rundell, eds., Springer, Vienna, 2000, pp. 147–154.
75. A. Louis and T. Schuster, *A novel filter design technique in 2D computerized tomography*, Inverse Problems, 12 (1996), pp. 685–696.
76. E. Macherauch and P. Müller, *Das* $\sin^2 \psi$*-Verfahren in der röntgenographischen Spannungsmessung*, Z. Angew. Physik, 13 (1961), pp. 305–312.
77. V. Maz'ja, *Sobolev spaces*, Springer, Heidelberg, 1985.
78. C. Müller, *Spherical Harmonics*, Lecture Notes in Mathematics, Springer, Berlin, New York, 1966.
79. F. Natterer, *Regularisierung schlecht gestellter Probleme durch Projektionsverfahren*, Numer. Math., 28 (1977), pp. 329–341.

80. ——, *The Mathematics of Computerized Tomography*, Wiley, Chichester, 1986.
81. ——, *The Mathematics of Computerized Tomography*, Wiley, Chichester, 1986.
82. ——, *Inverting the Attenuated Vectorial Radon transform*, J. Inv. Ill-Posed Prob., 13 (2005), pp. 93–101.
83. F. Natterer, M. Cheney, and B. Borden, *Resolution for radar and X-ray tomography*, Inverse Problems, 19 (2003), pp. S55–S66.
84. F. Natterer and F. Wübbeling, *Mathematical Methods in Image Reconstruction*, SIAM, Philadelphia, 2001.
85. A. Neubauer, *On Landweber iteration for nonlinear ill-posed problems in Hilbert scales*, Numer. Math., 85 (2000), pp. 309–328.
86. F. Noo, M. Defrise, and H. Kudo, *Exact reconstruction methods for multislice helical CT with a gantry tilt*, in Proc. VIIth Int. Conf. Fully 3D Reconstruction in Radiology and Nuclear Medicine, Y. Bizais, ed., 2003.
87. F. Noo, J. Pack, and D. Heuscher, *Exact helical reconstruction using native cone beam geometries*, Phys. Med. Biol., 48 (2003), pp. 3787–3818.
88. S. Norton, *Tomographic reconstruction of 2-D vector fields: Application to flow imaging*, Geophysics Journal, 97 (1988), pp. 161–168.
89. S. Norton and M. Linzer, *Ultrasonic reflectivity imaging in three dimensions: exact inverse scattering solutions for plane, cylindrical, and spherical apertures*, IEEE Trans. Biomed. Eng., 28 (1981), pp. 200–202.
90. N. Osman and J. Prince, *3D vector tomography on bounded domains*, Inverse Problems, 14 (1998), pp. 185–196.
91. V. Palamodov, *Reconstruction from limited data of arc means*, J. Fourier Anal. Appl., 6 (2000), pp. 26–42.
92. S. Patch, *Thermoacoustic tomography - consistency conditions and the partial scan problem*, Phys. Med. Biol., 49 (2004), pp. 2305–2315.
93. P.Liu, *Image reconstruction from photoacoustic pressure signals*, in Proceedings of the SPIE 1996, vol. 2681, 1996, pp. 285–296.
94. E. Quinto, *Singularities of the X-ray transform and limited data tomography in R^2 and R^3*, SIAM J. Math. Anal., 24 (1993), pp. 1215–1225.
95. J. Radon, *über die Bestimmung von Funktionen durch ihre Integralwerte längs gewisser Mannigfaltigkeiten*, Berichte Sächsische Akademie der Wissenschaften, Leipzig, 69 (1917), pp. 262–267.
96. A. Ramm, *Injectivity of the spherical mean operator*, C. R. ACad. Sci. Paris, I, 335 (2002), pp. 1033–1038.
97. F. Rashid-Farrokhi, K. Liu, C. Berenstein, and D. Walnut, *Wavelet based multiresolution local tomography*, IEEE Trans. Imag. Proc., 6 (1997), pp. 1412–1430.
98. A. Rieder, *On filter design principles in 2D computerized tomography*, in Radon Transforms and Tomography, E. Q. et al., ed., vol. 278 of Contemporary Mathematics, AMS Publications, 2001, pp. 207–226.
99. ——, *Keine Probleme mit Inversen Problemen (No Problems with Inverse Problems)*, Vieweg, Wiesbaden, 2003.
100. A. Rieder, R. Dietz, and T. Schuster, *Approximate inverse meets local tomography*, Math. Meth. Appl. Sci., 23 (2000), pp. 1373–1387.
101. A. Rieder and T. Schuster, *The approximate inverse in action with an application to computerized tomography*, SIAM J. Numer. Anal., 37 (2000), pp. 1909–1929.

102. ——, *The approximate inverse in action II: convergence and stability*, Mathematics of Computation, 72 (2003), pp. 1399–1415.
103. ——, *The approximate inverse in action III: 3D-Doppler tomography*, Numerische Mathematik, 97 (2004), pp. 353–378.
104. V. ROMANOV, *Integral Geometry and Inverse Problems for Hyperbolic Equations*, vol. 26 of Springer Tracts in Natural Philosophy, Springer Verlag, Berlin, 1974.
105. W. RUDIN, *Functional Analysis*, Tata McGraw-Hill, New York, 1973.
106. ——, *Real and Complex Analysis*, Mc Graw-Hill, New York, third ed., 1987.
107. F. SCHÖPFER, A. LOUIS, AND T. SCHUSTER, *Nonlinear iterative methods for linear ill-posed problems in Banach spaces*, Inverse Problems, 22 (2006), pp. 311–330.
108. L. SCHUMAKER, *Spline Functions: Basic Theory*, Wiley, Chichester, 1981.
109. T. SCHUSTER, *The 3D Doppler transform: elementary properties and computation of reconstruction kernels*, Inverse Problems, 16 (2000), pp. 701–723.
110. ——, *An efficient mollifier method for three-dimensional vector tomography: convergence analysis and implementation*, Inverse Problems, 17 (2001), pp. 739–766.
111. ——, *A stable inversion scheme for the Laplace transform using arbitrarily distributed data scanning points*, J. Inv. Ill-Posed Problems, 11 (2003), pp. 263–287.
112. ——, *Defect correction in vector field tomography: detecting the potential part using BEM and implementation of the method*, Inverse Problems, 21 (2005), pp. 75–91.
113. ——, *Error estimates for defect correction methods in Doppler tomography*, J. Inv. Ill-Posed Prob., 14 (2006), pp. 83–106.
114. T. SCHUSTER, J. PLÖGER, AND A. LOUIS, *Depth-resolved residual stress evaluation from X-ray diffraction measurement data using the approximate inverse method*, Z. Metallkd., 94 (2003), pp. 934–937.
115. T. SCHUSTER AND E. QUINTO, *On a regularization scheme for linear operators in distribution spaces with an application to the spherical Radon transform*, SIAM J. Appl. Math., 65 (2005), pp. 1369–1387.
116. V. SHARAFUTDINOV, *Integral Geometry of Tensor Fields*, VSP, Utrecht, 1994.
117. L. SHEPP AND B. LOGAN, *The Fourier reconstruction of a head section*, IEEE Trans. Nucl. Sci., NS-21 (1974), pp. 21–43.
118. H. SIELSCHOTT, *Measurement of horizontal flow in a large scale furnace using acoustic vector tomography*, Flow. Meas. Instrum., 8 (1997), pp. 191–197.
119. G. SPARR AND K. STRÅHLÉN, *Vector field tomography: an overview*, tech. report, Centre for Mathematical Sciences, Lund Institute of Technology, SE-221 00 Lund, Sweden, 1998. `http://www.lth.se/matematiklth/vision/publications.html#internal`.
120. G. SPARR, K. STRÅHLÉN, K. LINDSTRÖM, AND H. PERSSON, *Doppler tomography for vector fields*, Inverse Problems, 11 (1995), pp. 1051–1061.
121. F. STEFANI AND G. GERBETH, *Velocity reconstruction in conducting fluids from magnetic field and electric potential measurements*, Inverse Problems, 15 (1999), pp. 771–786.
122. ——, *On the uniqueness of velocity reconstruction in conducting fluids from measurements of induced electromagnetic fields*, Inverse Problems, 16 (2000), pp. 1–9.

123. E. STEIN, *Singular Integrals and Differentiability Properties of Functions*, Princeton University Press, Princeton, NJ, 1970.
124. K. STRÅHLÉN, *Reconstructions from Doppler Radon transforms*, tech. report, Lund Institute of Technology, Department of Mathematics, SE-221 00 Lund, Schweden, 1996.
125. A. TARANTOLA, *Inverse Problem Theory*, SIAM, Philadelphia, 2004.
126. H. TUY, *An inversion formula for cone-beam reconstruction*, SIAM J. Appl. Math., 43 (1983), pp. 546–552.
127. E. VAINBERG AND M. FAINGOIS, *Increasing the spatial resolution in computerized tomography*, in Problems in Tomographic Reconstruction, A. Alekseev, M. Lavrent'ev, and G. Preobrashensky, eds., Siberian Branch of the Academy of Science, USSR: Novosibirsk, 1985, pp. 28–35.
128. G. WANG, T. LIN, P. CHENG, AND D. SHINOZAKI, *A general cone-beam reconstruction algorithm*, IEEE Trans. Med. Imag., 12 (1993), pp. 486–496.
129. A. WERNSDÖRFER, *Complete reconstruction of three-dimensional vector fields*, in Proc. ECAPT 93, Karlsruhe, 1993.
130. H. WEYL, *The method of orthogonal projection in potential theory*, Duke Math. J., 7 (1940), pp. 411–444.
131. K. WINTERS AND D. ROUSEFF, *A filtered backprojection method for the tomographic reconstruction of fluid vorticity*, Inverse Problems, 6 (1990), pp. L33–L38.
132. M. XU, D. FENG, AND L.-H. WANG, *Exact frequency-domain reconstruction for thermoacoustic tomography-I: planar geometry*, IEEE Trans. Med. Imag., 21 (2002), pp. 823–828.
133. M. XU AND L.-H. WANG, *Time-domain reconstruction for thermoacoustic tomography in a spherical geometry*, IEEE Trans. Med. Imag., 21 (2002), pp. 814–822.
134. Y. XU, M. XU, AND L.-H. WANG, *Exact frequency-domain reconstruction for thermoacoustic tomography-II: spherical geometry*, IEEE Trans. Med. Imag., 21 (2002), pp. 829–833.

Index

Lecture Notes in Mathematics

For information about earlier volumes please contact your bookseller or Springer LNM Online archive: springerlink.com

Vol. 1711: W. Ricker, Operator Algebras Generated by Commuting Projections: A Vector Measure Approach (1999)
Vol. 1712: N. Schwartz, J. J. Madden, Semi-algebraic Function Rings and Reflectors of Partially Ordered Rings (1999)
Vol. 1713: F. Bethuel, G. Huisken, S. Müller, K. Steffen, Calculus of Variations and Geometric Evolution Problems. Cetraro, 1996. Editors: S. Hildebrandt, M. Struwe (1999)
Vol. 1714: O. Diekmann, R. Durrett, K. P. Hadeler, P. K. Maini, H. L. Smith, Mathematics Inspired by Biology. Martina Franca, 1997. Editors: V. Capasso, O. Diekmann (1999)
Vol. 1715: N. V. Krylov, M. Röckner, J. Zabczyk, Stochastic PDE's and Kolmogorov Equations in Infinite Dimensions. Cetraro, 1998. Editor: G. Da Prato (1999)
Vol. 1716: J. Coates, R. Greenberg, K. A. Ribet, K. Rubin, Arithmetic Theory of Elliptic Curves. Cetraro, 1997. Editor: C. Viola (1999)
Vol. 1717: J. Bertoin, F. Martinelli, Y. Peres, Lectures on Probability Theory and Statistics. Saint-Flour, 1997. Editor: P. Bernard (1999)
Vol. 1718: A. Eberle, Uniqueness and Non-Uniqueness of Semigroups Generated by Singular Diffusion Operators (1999)
Vol. 1719: K. R. Meyer, Periodic Solutions of the N-Body Problem (1999)
Vol. 1720: D. Elworthy, Y. Le Jan, X-M. Li, On the Geometry of Diffusion Operators and Stochastic Flows (1999)
Vol. 1721: A. Iarrobino, V. Kanev, Power Sums, Gorenstein Algebras, and Determinantal Loci (1999)
Vol. 1722: R. McCutcheon, Elemental Methods in Ergodic Ramsey Theory (1999)
Vol. 1723: J. P. Croisille, C. Lebeau, Diffraction by an Immersed Elastic Wedge (1999)
Vol. 1724: V. N. Kolokoltsov, Semiclassical Analysis for Diffusions and Stochastic Processes (2000)
Vol. 1725: D. A. Wolf-Gladrow, Lattice-Gas Cellular Automata and Lattice Boltzmann Models (2000)
Vol. 1726: V. Marić, Regular Variation and Differential Equations (2000)
Vol. 1727: P. Kravanja M. Van Barel, Computing the Zeros of Analytic Functions (2000)
Vol. 1728: K. Gatermann Computer Algebra Methods for Equivariant Dynamical Systems (2000)
Vol. 1729: J. Azéma, M. Émery, M. Ledoux, M. Yor (Eds.) Séminaire de Probabilités XXXIV (2000)
Vol. 1730: S. Graf, H. Luschgy, Foundations of Quantization for Probability Distributions (2000)
Vol. 1731: T. Hsu, Quilts: Central Extensions, Braid Actions, and Finite Groups (2000)
Vol. 1732: K. Keller, Invariant Factors, Julia Equivalences and the (Abstract) Mandelbrot Set (2000)
Vol. 1733: K. Ritter, Average-Case Analysis of Numerical Problems (2000)
Vol. 1734: M. Espedal, A. Fasano, A. Mikelić, Filtration in Porous Media and Industrial Applications. Cetraro 1998. Editor: A. Fasano. 2000.
Vol. 1735: D. Yafaev, Scattering Theory: Some Old and New Problems (2000)
Vol. 1736: B. O. Turesson, Nonlinear Potential Theory and Weighted Sobolev Spaces (2000)
Vol. 1737: S. Wakabayashi, Classical Microlocal Analysis in the Space of Hyperfunctions (2000)
Vol. 1738: M. Émery, A. Nemirovski, D. Voiculescu, Lectures on Probability Theory and Statistics (2000)
Vol. 1739: R. Burkard, P. Deuflhard, A. Jameson, J.-L. Lions, G. Strang, Computational Mathematics Driven by Industrial Problems. Martina Franca, 1999. Editors: V. Capasso, H. Engl, J. Periaux (2000)
Vol. 1740: B. Kawohl, O. Pironneau, L. Tartar, J.-P. Zolesio, Optimal Shape Design. Tróia, Portugal 1999. Editors: A. Cellina, A. Ornelas (2000)
Vol. 1741: E. Lombardi, Oscillatory Integrals and Phenomena Beyond all Algebraic Orders (2000)
Vol. 1742: A. Unterberger, Quantization and Non-holomorphic Modular Forms (2000)
Vol. 1743: L. Habermann, Riemannian Metrics of Constant Mass and Moduli Spaces of Conformal Structures (2000)
Vol. 1744: M. Kunze, Non-Smooth Dynamical Systems (2000)
Vol. 1745: V. D. Milman, G. Schechtman (Eds.), Geometric Aspects of Functional Analysis. Israel Seminar 1999-2000 (2000)
Vol. 1746: A. Degtyarev, I. Itenberg, V. Kharlamov, Real Enriques Surfaces (2000)
Vol. 1747: L. W. Christensen, Gorenstein Dimensions (2000)
Vol. 1748: M. Ruzicka, Electrorheological Fluids: Modeling and Mathematical Theory (2001)
Vol. 1749: M. Fuchs, G. Seregin, Variational Methods for Problems from Plasticity Theory and for Generalized Newtonian Fluids (2001)
Vol. 1750: B. Conrad, Grothendieck Duality and Base Change (2001)
Vol. 1751: N. J. Cutland, Loeb Measures in Practice: Recent Advances (2001)
Vol. 1752: Y. V. Nesterenko, P. Philippon, Introduction to Algebraic Independence Theory (2001)
Vol. 1753: A. I. Bobenko, U. Eitner, Painlevé Equations in the Differential Geometry of Surfaces (2001)
Vol. 1754: W. Bertram, The Geometry of Jordan and Lie Structures (2001)
Vol. 1755: J. Azéma, M. Émery, M. Ledoux, M. Yor (Eds.), Séminaire de Probabilités XXXV (2001)
Vol. 1756: P. E. Zhidkov, Korteweg de Vries and Nonlinear Schrödinger Equations: Qualitative Theory (2001)

Vol. 1757: R. R. Phelps, Lectures on Choquet's Theorem (2001)
Vol. 1758: N. Monod, Continuous Bounded Cohomology of Locally Compact Groups (2001)
Vol. 1759: Y. Abe, K. Kopfermann, Toroidal Groups (2001)
Vol. 1760: D. Filipović, Consistency Problems for Heath-Jarrow-Morton Interest Rate Models (2001)
Vol. 1761: C. Adelmann, The Decomposition of Primes in Torsion Point Fields (2001)
Vol. 1762: S. Cerrai, Second Order PDE's in Finite and Infinite Dimension (2001)
Vol. 1763: J.-L. Loday, A. Frabetti, F. Chapoton, F. Goichot, Dialgebras and Related Operads (2001)
Vol. 1764: A. Cannas da Silva, Lectures on Symplectic Geometry (2001)
Vol. 1765: T. Kerler, V. V. Lyubashenko, Non-Semisimple Topological Quantum Field Theories for 3-Manifolds with Corners (2001)
Vol. 1766: H. Hennion, L. Hervé, Limit Theorems for Markov Chains and Stochastic Properties of Dynamical Systems by Quasi-Compactness (2001)
Vol. 1767: J. Xiao, Holomorphic Q Classes (2001)
Vol. 1768: M. J. Pflaum, Analytic and Geometric Study of Stratified Spaces (2001)
Vol. 1769: M. Alberich-Carramiñana, Geometry of the Plane Cremona Maps (2002)
Vol. 1770: H. Gluesing-Luerssen, Linear Delay-Differential Systems with Commensurate Delays: An Algebraic Approach (2002)
Vol. 1771: M. Émery, M. Yor (Eds.), Séminaire de Probabilités 1967-1980. A Selection in Martingale Theory (2002)
Vol. 1772: F. Burstall, D. Ferus, K. Leschke, F. Pedit, U. Pinkall, Conformal Geometry of Surfaces in S^4 (2002)
Vol. 1773: Z. Arad, M. Muzychuk, Standard Integral Table Algebras Generated by a Non-real Element of Small Degree (2002)
Vol. 1774: V. Runde, Lectures on Amenability (2002)
Vol. 1775: W. H. Meeks, A. Ros, H. Rosenberg, The Global Theory of Minimal Surfaces in Flat Spaces. Martina Franca 1999. Editor: G. P. Pirola (2002)
Vol. 1776: K. Behrend, C. Gomez, V. Tarasov, G. Tian, Quantum Comohology. Cetraro 1997. Editors: P. de Bartolomeis, B. Dubrovin, C. Reina (2002)
Vol. 1777: E. García-Río, D. N. Kupeli, R. Vázquez-Lorenzo, Osserman Manifolds in Semi-Riemannian Geometry (2002)
Vol. 1778: H. Kiechle, Theory of K-Loops (2002)
Vol. 1779: I. Chueshov, Monotone Random Systems (2002)
Vol. 1780: J. H. Bruinier, Borcherds Products on O(2,1) and Chern Classes of Heegner Divisors (2002)
Vol. 1781: E. Bolthausen, E. Perkins, A. van der Vaart, Lectures on Probability Theory and Statistics. Ecole d' Eté de Probabilités de Saint-Flour XXIX-1999. Editor: P. Bernard (2002)
Vol. 1782: C.-H. Chu, A. T.-M. Lau, Harmonic Functions on Groups and Fourier Algebras (2002)
Vol. 1783: L. Grüne, Asymptotic Behavior of Dynamical and Control Systems under Perturbation and Discretization (2002)
Vol. 1784: L. H. Eliasson, S. B. Kuksin, S. Marmi, J.-C. Yoccoz, Dynamical Systems and Small Divisors. Cetraro, Italy 1998. Editors: S. Marmi, J.-C. Yoccoz (2002)
Vol. 1785: J. Arias de Reyna, Pointwise Convergence of Fourier Series (2002)
Vol. 1786: S. D. Cutkosky, Monomialization of Morphisms from 3-Folds to Surfaces (2002)
Vol. 1787: S. Caenepeel, G. Militaru, S. Zhu, Frobenius and Separable Functors for Generalized Module Categories and Nonlinear Equations (2002)
Vol. 1788: A. Vasil'ev, Moduli of Families of Curves for Conformal and Quasiconformal Mappings (2002)
Vol. 1789: Y. Sommerhäuser, Yetter-Drinfel'd Hopf algebras over groups of prime order (2002)
Vol. 1790: X. Zhan, Matrix Inequalities (2002)
Vol. 1791: M. Knebusch, D. Zhang, Manis Valuations and Prüfer Extensions I: A new Chapter in Commutative Algebra (2002)
Vol. 1792: D. D. Ang, R. Gorenflo, V. K. Le, D. D. Trong, Moment Theory and Some Inverse Problems in Potential Theory and Heat Conduction (2002)
Vol. 1793: J. Cortés Monforte, Geometric, Control and Numerical Aspects of Nonholonomic Systems (2002)
Vol. 1794: N. Pytheas Fogg, Substitution in Dynamics, Arithmetics and Combinatorics. Editors: V. Berthé, S. Ferenczi, C. Mauduit, A. Siegel (2002)
Vol. 1795: H. Li, Filtered-Graded Transfer in Using Non-commutative Gröbner Bases (2002)
Vol. 1796: J.M. Melenk, hp-Finite Element Methods for Singular Perturbations (2002)
Vol. 1797: B. Schmidt, Characters and Cyclotomic Fields in Finite Geometry (2002)
Vol. 1798: W.M. Oliva, Geometric Mechanics (2002)
Vol. 1799: H. Pajot, Analytic Capacity, Rectifiability, Menger Curvature and the Cauchy Integral (2002)
Vol. 1800: O. Gabber, L. Ramero, Almost Ring Theory (2003)
Vol. 1801: J. Azéma, M. Émery, M. Ledoux, M. Yor (Eds.), Séminaire de Probabilités XXXVI (2003)
Vol. 1802: V. Capasso, E. Merzbach, B. G. Ivanoff, M. Dozzi, R. Dalang, T. Mountford, Topics in Spatial Stochastic Processes. Martina Franca, Italy 2001. Editor: E. Merzbach (2003)
Vol. 1803: G. Dolzmann, Variational Methods for Crystalline Microstructure – Analysis and Computation (2003)
Vol. 1804: I. Cherednik, Ya. Markov, R. Howe, G. Lusztig, Iwahori-Hecke Algebras and their Representation Theory. Martina Franca, Italy 1999. Editors: V. Baldoni, D. Barbasch (2003)
Vol. 1805: F. Cao, Geometric Curve Evolution and Image Processing (2003)
Vol. 1806: H. Broer, I. Hoveijn. G. Lunther, G. Vegter, Bifurcations in Hamiltonian Systems. Computing Singularities by Gröbner Bases (2003)
Vol. 1807: V. D. Milman, G. Schechtman (Eds.), Geometric Aspects of Functional Analysis. Israel Seminar 2000-2002 (2003)
Vol. 1808: W. Schindler, Measures with Symmetry Properties (2003)
Vol. 1809: O. Steinbach, Stability Estimates for Hybrid Coupled Domain Decomposition Methods (2003)
Vol. 1810: J. Wengenroth, Derived Functors in Functional Analysis (2003)
Vol. 1811: J. Stevens, Deformations of Singularities (2003)
Vol. 1812: L. Ambrosio, K. Deckelnick, G. Dziuk, M. Mimura, V. A. Solonnikov, H. M. Soner, Mathematical Aspects of Evolving Interfaces. Madeira, Funchal, Portugal 2000. Editors: P. Colli, J. F. Rodrigues (2003)
Vol. 1813: L. Ambrosio, L. A. Caffarelli, Y. Brenier, G. Buttazzo, C. Villani, Optimal Transportation and its

Applications. Martina Franca, Italy 2001. Editors: L. A. Caffarelli, S. Salsa (2003)
Vol. 1814: P. Bank, F. Baudoin, H. Föllmer, L.C.G. Rogers, M. Soner, N. Touzi, Paris-Princeton Lectures on Mathematical Finance 2002 (2003)
Vol. 1815: A. M. Vershik (Ed.), Asymptotic Combinatorics with Applications to Mathematical Physics. St. Petersburg, Russia 2001 (2003)
Vol. 1816: S. Albeverio, W. Schachermayer, M. Talagrand, Lectures on Probability Theory and Statistics. Ecole d'Eté de Probabilités de Saint-Flour XXX-2000. Editor: P. Bernard (2003)
Vol. 1817: E. Koelink, W. Van Assche (Eds.), Orthogonal Polynomials and Special Functions. Leuven 2002 (2003)
Vol. 1818: M. Bildhauer, Convex Variational Problems with Linear, nearly Linear and/or Anisotropic Growth Conditions (2003)
Vol. 1819: D. Masser, Yu. V. Nesterenko, H. P. Schlickewei, W. M. Schmidt, M. Waldschmidt, Diophantine Approximation. Cetraro, Italy 2000. Editors: F. Amoroso, U. Zannier (2003)
Vol. 1820: F. Hiai, H. Kosaki, Means of Hilbert Space Operators (2003)
Vol. 1821: S. Teufel, Adiabatic Perturbation Theory in Quantum Dynamics (2003)
Vol. 1822: S.-N. Chow, R. Conti, R. Johnson, J. Mallet-Paret, R. Nussbaum, Dynamical Systems. Cetraro, Italy 2000. Editors: J. W. Macki, P. Zecca (2003)
Vol. 1823: A. M. Anile, W. Allegretto, C. Ringhofer, Mathematical Problems in Semiconductor Physics. Cetraro, Italy 1998. Editor: A. M. Anile (2003)
Vol. 1824: J. A. Navarro González, J. B. Sancho de Salas, $\mathscr{C}^\infty$ – Differentiable Spaces (2003)
Vol. 1825: J. H. Bramble, A. Cohen, W. Dahmen, Multiscale Problems and Methods in Numerical Simulations, Martina Franca, Italy 2001. Editor: C. Canuto (2003)
Vol. 1826: K. Dohmen, Improved Bonferroni Inequalities via Abstract Tubes. Inequalities and Identities of Inclusion-Exclusion Type. VIII, 113 p, 2003.
Vol. 1827: K. M. Pilgrim, Combinations of Complex Dynamical Systems. IX, 118 p, 2003.
Vol. 1828: D. J. Green, Gröbner Bases and the Computation of Group Cohomology. XII, 138 p, 2003.
Vol. 1829: E. Altman, B. Gaujal, A. Hordijk, Discrete-Event Control of Stochastic Networks: Multimodularity and Regularity. XIV, 313 p, 2003.
Vol. 1830: M. I. Gil', Operator Functions and Localization of Spectra. XIV, 256 p, 2003.
Vol. 1831: A. Connes, J. Cuntz, E. Guentner, N. Higson, J. E. Kaminker, Noncommutative Geometry, Martina Franca, Italy 2002. Editors: S. Doplicher, L. Longo (2004)
Vol. 1832: J. Azéma, M. Émery, M. Ledoux, M. Yor (Eds.), Séminaire de Probabilités XXXVII (2003)
Vol. 1833: D.-Q. Jiang, M. Qian, M.-P. Qian, Mathematical Theory of Nonequilibrium Steady States. On the Frontier of Probability and Dynamical Systems. IX, 280 p, 2004.
Vol. 1834: Yo. Yomdin, G. Comte, Tame Geometry with Application in Smooth Analysis. VIII, 186 p, 2004.
Vol. 1835: O.T. Izhboldin, B. Kahn, N.A. Karpenko, A. Vishik, Geometric Methods in the Algebraic Theory of Quadratic Forms. Summer School, Lens, 2000. Editor: J.-P. Tignol (2004)
Vol. 1836: C. Năstăsescu, F. Van Oystaeyen, Methods of Graded Rings. XIII, 304 p, 2004.
Vol. 1837: S. Tavaré, O. Zeitouni, Lectures on Probability Theory and Statistics. Ecole d'Eté de Probabilités de Saint-Flour XXXI-2001. Editor: J. Picard (2004)
Vol. 1838: A.J. Ganesh, N.W. O'Connell, D.J. Wischik, Big Queues. XII, 254 p, 2004.
Vol. 1839: R. Gohm, Noncommutative Stationary Processes. VIII, 170 p, 2004.
Vol. 1840: B. Tsirelson, W. Werner, Lectures on Probability Theory and Statistics. Ecole d'Eté de Probabilités de Saint-Flour XXXII-2002. Editor: J. Picard (2004)
Vol. 1841: W. Reichel, Uniqueness Theorems for Variational Problems by the Method of Transformation Groups (2004)
Vol. 1842: T. Johnsen, A. L. Knutsen, K_3 Projective Models in Scrolls (2004)
Vol. 1843: B. Jefferies, Spectral Properties of Noncommuting Operators (2004)
Vol. 1844: K.F. Siburg, The Principle of Least Action in Geometry and Dynamics (2004)
Vol. 1845: Min Ho Lee, Mixed Automorphic Forms, Torus Bundles, and Jacobi Forms (2004)
Vol. 1846: H. Ammari, H. Kang, Reconstruction of Small Inhomogeneities from Boundary Measurements (2004)
Vol. 1847: T.R. Bielecki, T. Björk, M. Jeanblanc, M. Rutkowski, J.A. Scheinkman, W. Xiong, Paris-Princeton Lectures on Mathematical Finance 2003 (2004)
Vol. 1848: M. Abate, J. E. Fornaess, X. Huang, J. P. Rosay, A. Tumanov, Real Methods in Complex and CR Geometry, Martina Franca, Italy 2002. Editors: D. Zaitsev, G. Zampieri (2004)
Vol. 1849: Martin L. Brown, Heegner Modules and Elliptic Curves (2004)
Vol. 1850: V. D. Milman, G. Schechtman (Eds.), Geometric Aspects of Functional Analysis. Israel Seminar 2002-2003 (2004)
Vol. 1851: O. Catoni, Statistical Learning Theory and Stochastic Optimization (2004)
Vol. 1852: A.S. Kechris, B.D. Miller, Topics in Orbit Equivalence (2004)
Vol. 1853: Ch. Favre, M. Jonsson, The Valuative Tree (2004)
Vol. 1854: O. Saeki, Topology of Singular Fibers of Differential Maps (2004)
Vol. 1855: G. Da Prato, P.C. Kunstmann, I. Lasiecka, A. Lunardi, R. Schnaubelt, L. Weis, Functional Analytic Methods for Evolution Equations. Editors: M. Iannelli, R. Nagel, S. Piazzera (2004)
Vol. 1856: K. Back, T.R. Bielecki, C. Hipp, S. Peng, W. Schachermayer, Stochastic Methods in Finance, Bressanone/Brixen, Italy, 2003. Editors: M. Fritelli, W. Runggaldier (2004)
Vol. 1857: M. Émery, M. Ledoux, M. Yor (Eds.), Séminaire de Probabilités XXXVIII (2005)
Vol. 1858: A.S. Cherny, H.-J. Engelbert, Singular Stochastic Differential Equations (2005)
Vol. 1859: E. Letellier, Fourier Transforms of Invariant Functions on Finite Reductive Lie Algebras (2005)
Vol. 1860: A. Borisyuk, G.B. Ermentrout, A. Friedman, D. Terman, Tutorials in Mathematical Biosciences I. Mathematical Neurosciences (2005)
Vol. 1861: G. Benettin, J. Henrard, S. Kuksin, Hamiltonian Dynamics – Theory and Applications, Cetraro, Italy, 1999. Editor: A. Giorgilli (2005)
Vol. 1862: B. Helffer, F. Nier, Hypoelliptic Estimates and Spectral Theory for Fokker-Planck Operators and Witten Laplacians (2005)

Vol. 1863: H. Führ, Abstract Harmonic Analysis of Continuous Wavelet Transforms (2005)
Vol. 1864: K. Efstathiou, Metamorphoses of Hamiltonian Systems with Symmetries (2005)
Vol. 1865: D. Applebaum, B.V. R. Bhat, J. Kustermans, J. M. Lindsay, Quantum Independent Increment Processes I. From Classical Probability to Quantum Stochastic Calculus. Editors: M. Schürmann, U. Franz (2005)
Vol. 1866: O.E. Barndorff-Nielsen, U. Franz, R. Gohm, B. Kümmerer, S. Thorbjønsen, Quantum Independent Increment Processes II. Structure of Quantum Lévy Processes, Classical Probability, and Physics. Editors: M. Schürmann, U. Franz, (2005)
Vol. 1867: J. Sneyd (Ed.), Tutorials in Mathematical Biosciences II. Mathematical Modeling of Calcium Dynamics and Signal Transduction. (2005)
Vol. 1868: J. Jorgenson, S. Lang, $Pos_n(R)$ and Eisenstein Series. (2005)
Vol. 1869: A. Dembo, T. Funaki, Lectures on Probability Theory and Statistics. Ecole d'Eté de Probabilités de Saint-Flour XXXIII-2003. Editor: J. Picard (2005)
Vol. 1870: V.I. Gurariy, W. Lusky, Geometry of Müntz Spaces and Related Questions. (2005)
Vol. 1871: P. Constantin, G. Gallavotti, A.V. Kazhikhov, Y. Meyer, S. Ukai, Mathematical Foundation of Turbulent Viscous Flows, Martina Franca, Italy, 2003. Editors: M. Cannone, T. Miyakawa (2006)
Vol. 1872: A. Friedman (Ed.), Tutorials in Mathematical Biosciences III. Cell Cycle, Proliferation, and Cancer (2006)
Vol. 1873: R. Mansuy, M. Yor, Random Times and Enlargements of Filtrations in a Brownian Setting (2006)
Vol. 1874: M. Yor, M. Émery (Eds.), In Memoriam Paul-André Meyer - Séminaire de probabilités XXXIX (2006)
Vol. 1875: J. Pitman, Combinatorial Stochastic Processes. Ecole d'Eté de Probabilités de Saint-Flour XXXII-2002. Editor: J. Picard (2006)
Vol. 1876: H. Herrlich, Axiom of Choice (2006)
Vol. 1877: J. Steuding, Value Distributions of L-Functions (2007)
Vol. 1878: R. Cerf, The Wulff Crystal in Ising and Percolation Models, Ecole d'Eté de Probabilités de Saint-Flour XXXIV-2004. Editor: Jean Picard (2006)
Vol. 1879: G. Slade, The Lace Expansion and its Applications, Ecole d'Eté de Probabilités de Saint-Flour XXXIV-2004. Editor: Jean Picard (2006)
Vol. 1880: S. Attal, A. Joye, C.-A. Pillet, Open Quantum Systems I, The Hamiltonian Approach (2006)
Vol. 1881: S. Attal, A. Joye, C.-A. Pillet, Open Quantum Systems II, The Markovian Approach (2006)
Vol. 1882: S. Attal, A. Joye, C.-A. Pillet, Open Quantum Systems III, Recent Developments (2006)
Vol. 1883: W. Van Assche, F. Marcellàn (Eds.), Orthogonal Polynomials and Special Functions, Computation and Application (2006)
Vol. 1884: N. Hayashi, E.I. Kaikina, P.I. Naumkin, I.A. Shishmarev, Asymptotics for Dissipative Nonlinear Equations (2006)
Vol. 1885: A. Telcs, The Art of Random Walks (2006)
Vol. 1886: S. Takamura, Splitting Deformations of Degenerations of Complex Curves (2006)
Vol. 1887: K. Habermann, L. Habermann, Introduction to Symplectic Dirac Operators (2006)
Vol. 1888: J. van der Hoeven, Transseries and Real Differential Algebra (2006)
Vol. 1889: G. Osipenko, Dynamical Systems, Graphs, and Algorithms (2006)
Vol. 1890: M. Bunge, J. Funk, Singular Coverings of Toposes (2006)
Vol. 1891: J.B. Friedlander, D.R. Heath-Brown, H. Iwaniec, J. Kaczorowski, Analytic Number Theory, Cetraro, Italy, 2002. Editors: A. Perelli, C. Viola (2006)
Vol. 1892: A. Baddeley, I. Bárány, R. Schneider, W. Weil, Stochastic Geometry, Martina Franca, Italy, 2004. Editor: W. Weil (2007)
Vol. 1893: H. Hanßmann, Local and Semi-Local Bifurcations in Hamiltonian Dynamical Systems, Results and Examples (2007)
Vol. 1894: C.W. Groetsch, Stable Approximate Evaluation of Unbounded Operators (2007)
Vol. 1895: L. Molnár, Selected Preserver Problems on Algebraic Structures of Linear Operators and on Function Spaces (2007)
Vol. 1896: P. Massart, Concentration Inequalities and Model Selection, Ecole d'Eté de Probabilités de Saint-Flour XXXIII-2003. Editor: J. Picard (2007)
Vol. 1897: R. Doney, Fluctuation Theory for Lévy Processes, Ecole d'Eté de Probabilités de Saint-Flour XXXV-2005. Editor: J. Picard (2007)
Vol. 1898: H.R. Beyer, Beyond Partial Differential Equations, On linear and Quasi-Linear Abstract Hyperbolic Evolution Equations (2007)
Vol. 1899: Séminaire de Probabilités XL. Editors: C. Donati-Martin, M. Émery, A. Rouault, C. Stricker (2007)
Vol. 1900: E. Bolthausen, A. Bovier (Eds.), Spin Glasses (2007)
Vol. 1901: O. Wittenberg, Intersections de deux quadriques et pinceaux de courbes de genre 1, Intersections of Two Quadrics and Pencils of Curves of Genus 1 (2007)
Vol. 1902: A. Isaev, Lectures on the Automorphism Groups of Kobayashi-Hyperbolic Manifolds (2007)
Vol. 1903: G. Kresin, V. Maz'ya, Sharp Real-Part Theorems (2007)
Vol. 1904: P. Giesl, Construction of Global Lyapunov Functions Using Radial Basis Functions (2007)
Vol. 1905: C. Prévôt, M. Röckner, A Concise Course on Stochastic Partial Differential Equations (2007)
Vol. 1906: T. Schuster, The Method of Approximate Inverse: Theory and Applications (2007)

Recent Reprints and New Editions

Vol. 1618: G. Pisier, Similarity Problems and Completely Bounded Maps. 1995 – 2nd exp. edition (2001)
Vol. 1629: J.D. Moore, Lectures on Seiberg-Witten Invariants. 1997 – 2nd edition (2001)
Vol. 1638: P. Vanhaecke, Integrable Systems in the realm of Algebraic Geometry. 1996 – 2nd edition (2001)
Vol. 1702: J. Ma, J. Yong, Forward-Backward Stochastic Differential Equations and their Applications. 1999 – Corr. 3rd printing (2007)
Vol. 830: J.A. Green, Polynomial Representations of GL_n, with an Appendix on Schensted Correspondence and Littelmann Paths by K. Erdmann, J.A. Green and M. Schocker 1980 – 2nd corr. and augmented edition (2007)